Graphing Calculator Manual

HOLT, RINEHART AND WINSTON
A Harcourt Education Company
Orlando • Austin • New York • San Diego • Toronto • London

Printed in the United States of America

ISBN 0-03-041652-3

1 2 3 4 018 08 07 06 05

Table of Contents

How To Use This Manual

Calculators on the market today are extremely sophisticated and incorporate many different features. This advanced functionality allows calculator users to solve problems in more than one way. The methods explained in this manual are not the only way to approach each topic; there are other approaches that you may discover on your own that will work as well.

This manual presents a basic foundation for the use of 7 different calculators. All of these calculators are reasonable choices to do the calculator problems in the textbook. There are other calculators not discussed in this manual that also are suitable for this material.

To use this manual effectively, you should begin by reading Chapter 1, *Basic Calculator Topics*. This chapter discusses topics pertaining to the use of technology in a mathematics class, with an emphasis on graphing calculators. It contains useful material on how to use your calculator, how to get a clear and useful graph, and when to watch for technology errors that may occur in some calculations.

Chapter 2, *Calculator Notes and Problems*, explains procedures that are applicable to a wide range of calculators. The narrative material parallels and refers to the problems in *Contemporary Precalculus: A Graphing Approach 3rd edition*, by Thomas W. Hungerford. The problems discussed are suitable for technology solutions. Included is a review of the pertinent mathematics needed to do these problems. Numerous examples are presented, as well as problems with detailed explanations on the graphing calculator solutions. Each chapter in the textbook is covered by a separate section in Chapter 2. The title of the corresponding chapter is used as the head of each section.

Chapters 3 through 9 provide directions and keystrokes for 7 different calculators currently available and suitable for doing the problems in this manual and from the Hungerford textbook. Each calculator method shows the *specific keystrokes* needed to enter the functions used in the text. Before you begin using your calculator for this course, we recommend a thorough review of the chapter that applies to your calculator.

When you are comfortable with the introductory methods explained in this manual, the calculator manual that accompanies your calculator will be easier to understand. Then you can explore more advanced features of your calculator.

Chapter 1 - BASIC CALCULATOR TOPICS

Technology Pitfalls

Technology gives us powerful tools to study mathematics. Many practical applied problems would be very difficult and time consuming or even impossible to solve with pencil and paper. However, technology has its limits and must be used carefully. Here is a brief discussion of some of the big pitfalls.

1. *Round- off and Cancellation Errors*

The number of significant digits in a decimal is the number of digits between the first and last nonzero digit. Calculators can handle only a specified number of digits. They either round off the number to the specified number of digits or truncate (drop off) the digits they can not handle. When many numbers are used in a calculation, the cumulative effect of many small round-off errors can be large.

Round-off errors can lead to cancellation errors; these errors occur in computations of small differences between large numbers. Cancellation errors can produce an error of magnitude 10. Numbers that are different will be the same after rounding

2. *Overflow and Underflow Errors*

Each calculator can handle only a given number of digits. If a number is larger than the calculator can store, the error is called overflow. If the number is too small for the calculator to store, the error is called underflow. When a small number is divided by a large number the answer may be smaller than the calculator can store. The number is stored as zero. This problem occurs on any piece of technology, but it is more noticeable on calculators that store a small number of digits.

3. *Numeric Calculation Errors*

The newer calculators have numeric solve algorithms programmed into the calculator. These algorithms have been programmed to perform a specified number of steps to do the calculation. If a particular problem needs more steps to get a reasonable approximation, the approximate answer may not have the accuracy required by the problem. In some algorithms, you can change the number of steps in the algorithm, but this is not always true.

4. *Hidden Behavior of Graphs*

The screen of a calculator is made up of pixels or dots which are turned on or off by the calculator to display the picture. The picture that is shown on the calculator screen is a finite set of dots while the true graph consists of an infinite number of points. The picture that you see is an approximation to the real thing. The screen is finite in size so the entire graph of the function cannot be seen in any window. Watch for the following problems.

a) The graph may lie in an area outside the viewing window.

b) The graph may have blank spots or jumps due to round-off, cancellation, overflow or underflow errors.

c) Zooming out may lose all of the local properties.

d) Zooming in may lose all of the global properties of the function. Zooming in too far often calculates numbers that are too small for the calculator to handle.

e) Extra lines appear when points are connected in the viewing window that are not connected on the true graph. This frequently happens when the graph has vertical asymptotes.

What You See May Not Be What You Have!

Exact Answer or Approximate Answer

Exact Answer

When the Algebraic solution to an equation is an integer or a rational number (Fraction), then the answer is exact.

ex. $3x = 7$ $\quad$ $4x = 2$

$x = \frac{7}{3}$ $\quad$ $x = \frac{1}{2}$

When the Algebraic solution to an equation is a function of an integer or a rational number (Fraction), then the answer is exact.

ex. $x^2 - 2 = 0$ $\quad$ $10^x = 2$

$x = \pm\sqrt{2}$ $\quad$ $x = \log(2)$

A calculator with CAS (Computer Algebra System) can display exact answers. These calculators at present are the TI 89, TI 92, HP 38, HP 48G and Casio fx 2.

Approximations

The other graphing calculators are numeric. They only calculate with numbers, they do not do symbolic manipulations. These calculators display the exact answers with a decimal representation. These decimal representations, unless they terminate or repeat, are approximations. The decimal representation goes on forever and when we round or truncate the result is a decimal approximation.

Ex. $\sqrt{2} \approx 1.414213562...$ $\quad$ $\log(2) \approx .3010299957...$

The algorithms used to compute roots and other functions give decimal approximations to the exact answer. If all of the digits displayed are used for the approximation, the approximation will check as the computation is at the limit of accuracy for the calculator. If the decimal approximation of the answer is rounded or truncated, the approximation will not check.

Every exact answer will check.

Ex. $(\sqrt{2})^2 - 2 = 0$ $\qquad 10^{\log(2)} = 2$

$(-\sqrt{2})^2 - 2 = 0$ $\qquad 10^{0.301} = 1.99986187 \neq 2$

$(1.414)^2 - 2 = -.000604 \neq 0$

This manual will consider approximations rounded to 2 or 3 decimal places.

Entering Algebraic Functions

Functions consist of an input, a rule and an output. Calculators are function machines. Type in the rule, then the input, the calculator displays the output. Before calculators, functions were read from tables. Find the page for the function, look for the input, read the output from the table. When a function is entered into an algebraic graphing calculator, the same procedure is followed. To compute a square root: Enter $\sqrt{}$ (the function), enter the input 3, Enter (or Execute) to get the output: $\sqrt{3} = 1.732...$

Suppose the square root of more than one number is needed. Enter $\sqrt{}\, x$. On the calculator, the x, or any number, does not go into the square root symbol, it is typed after it. How does the calculator know how much goes into the square root calculation? What if the problem was $\sqrt{x^2 + 2x}$? To tell the calculator that the entire input is $x^2 + 2x$, this expression , called the argument, is put in parenthesis. So on the calculator screen, $\sqrt{}\,(x^2 + 2x)$. Some calculators like the TI 83 force the parenthesis, otherwise you <u>**MUST**</u> remember to put them in.

These parenthesis around the argument must be used for all of the functions and when entering fractions.

ex. $\sin(x^2 + 2x)$

$\log(x^2 + 2x)$

$(x^2 + 2x)^2$

$abs(x^2 + 2x)$

$e^{(x^2+2x)}$

$\frac{(x^2+2x)}{(x^3+5)}$

Now the calculator knows exactly what it is to do.

The Rule of Three

There is more than one way to represent an algebraic function. It can be represented by an equation $y = 2x^2 - 4x + 7$

1. The equation is the <u>Algebraic</u> representation.

This function can be entered into the $y =$ menu of a graphing calculator, then a graph or picture of the function can be drawn. This graph will only show part of the function as the graph screen is finite in size, and this function has a domain of all real numbers.

2. The graph is the Graphical representation.

   Using the TABLE option in most graphing calculators (not the TI 81 or TI 85), a table of values for the function can be constructed. All of the possible input values in the domain can not be shown, but selective ones can be displayed.

3. The table is the Numeric representation of the function.

All of the representations of the Rule of Three can be displayed on the calculator. Graphical and Numeric representations occur more frequently in application problems where real data is used.

The Calculator Window

Pixels

The screen of a calculator is made up of pixels, electronic dots that are turned on and off by the calculator to display the picture. Each pixel is assigned a size, determined by the scale sizes in the window settings. The numeric value of a pixel = (Xmax – Xmin) ÷ (the number of pixels on the length of the calculator screen).

True Shape

It is convenient to have the graphs displayed on the calculator be their true shape. For example, circles should be round, not elliptical.

Pictures are their true shape if the X and Y values of the pixel are the same number. Every calculator has a window where each pixel has a value of 0.1 in both the X and Y directions. It is the Zoom Decimal or Initialize window. On this window, pictures are their correct shape. When the dimensions on the X-axis are determined, set the pixels on the Y-axis the same size and the graph will be its True Shape. Pixel size = (Ymax – Ymin) ÷ (the number of pixels on the height of the calculator screen) = (Xmax – Xmin) ÷ (the number of pixels on the width of the calculator screen).

Friendly Window

It is also convenient to have the X-coordinates when Trace is activated to have a "nice" size. say 1 unit or 0.1 unit. Any window where the X-values of the pixels are nice decimals like 10, 1, 0.1, 0.2, 0.25, 0.5 etc. is called a Friendly Window. This happens when Xmax – Xmin is a factor or multiple of the number of pixels on the length of the calculator screen (decimal factors are allowed).

For example: Suppose there are 94 pixels on the length of the screen and 62 pixels on the height. Then any window with Xmax – Xmin = k(94) will be Friendly.

The Decimal window has Xmin = -4.7, Xmax = 4.7, this is a Friendly Window with k = .1.

Consider the window with Xmin = 20, Xmax = 29.4, as this window also has Xmax – Xmin = 9.4.

It is also a Friendly Window with k = .1. Translating a Friendly Window is still a Friendly Window.

To make this example a window where graphs are their true shape, pick Ymax –Ymin = 0.1 (63) = 6.3

ex: $y = \frac{1}{x-25}$ Find a Friendly Window showing the discontinuity.

The discontinuity is at x = 25. This is the value of x that makes the denominator = 0, so the function is not defined at this point. First set the decimal window. Get a Friendly Window with k = 0.1, but center the screen at x = 25.

For a screen with 94 horizontal pixels, this means Xmin $= 25 - 4.7 = 20.3$ and Xmax $= 25 + 4.7 = 29.7$. Graph the function.

For a screen with 126 horizontal pixels, this means Xmin $= 25 - 6.3 = 18.7$ and Xmax $= 25 + 6.3 = 31.3$. Graph the function.

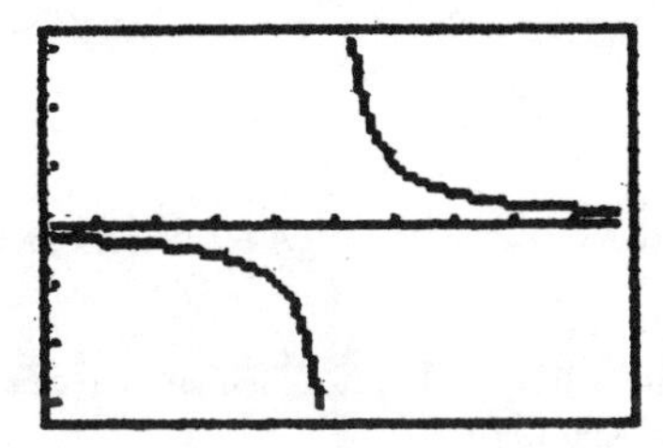

This is a complete graph on a Friendly Window. Because the Window was set to the Decimal Window and then translated, the units on the Y axis are the same size as the units on the X axis, so the graph is the true shape.

By the Big-Little Rule, $1 \div Big = Little$ so there is a horizontal asymptote at $y = 0$. On some calculators a vertical line is shown at $x = 25$. This is a technology error and should be ignored. The calculator has connected the points on both sides of the asymptote.

Complete Graph

A Complete Graph is one that displays all of the important features of the graph: the x-intercepts, the y-intercepts, local maximums, local minimums and end behavior. Sometimes a Complete Graph is more important than a graph which displays the true shape of the graph.

Different calculators have different size screens. Here is a list of the sizes of the current models of Graphing Calculators.

Calculator		Pixels in Length	Pixels in Height
T.I.	82	94	62
T.I.	83	94	62
T.I.	85	126	62
T.I.	86	126	62
T.I.	89	158	76
T.I.	92	238	102
Casio	9850	127	63
Casio	fx.2	127	63
H.P.	38	130	63
H.P.	48G	130	63
Sharp	9600	126	62

Hints to find a Friendly Window

1. Write a list of some multiples and factors of the number of pixels on the length of the calculator screen.
2. Determine x values for the important characteristics of the function: intercepts,asymptotes, local max. and min.
3. Compute the difference between the largest and smallest x values from part 2. Find an interval from part 1 that is longer than this. Use this interval as Xmax – Xmin, center so that all the values in part 2 are on the interval.
4. Graph the function.
5. If the height of the graph is not correct try to autoscale. This may or may not give you a good graph.
6. If you need to resize the Y axis, follow steps 1 through 4 using the number of pixels for the height of the screen.

Copying a Calculator Graph to Paper

1. Graph the function on the calculator, set the window so that a complete graph is displayed. Try to draw it in a Friendly Window.
2. Draw the window on paper. Draw the x-axis and the y-axis if they are in the window to be copied. Label the endpoints of the window: Xmin, Xmax, Ymin, Ymax and the scale marks on the axis.
3. Look at the graph on the calculator. Does it have any local maxima, local minima or asymptotes? Lightly sketch these on the window on the paper.

4. Press **[TRACE]** Mark 4 to 6 values from the Calculator graph onto the paper. Include any maximum or minimum values. Join the points to look like the calculator graph.

Problem: $f(x) = \frac{2x}{\sqrt{x^2+x+1}}$ Use the Decimal Window and copy the graph.

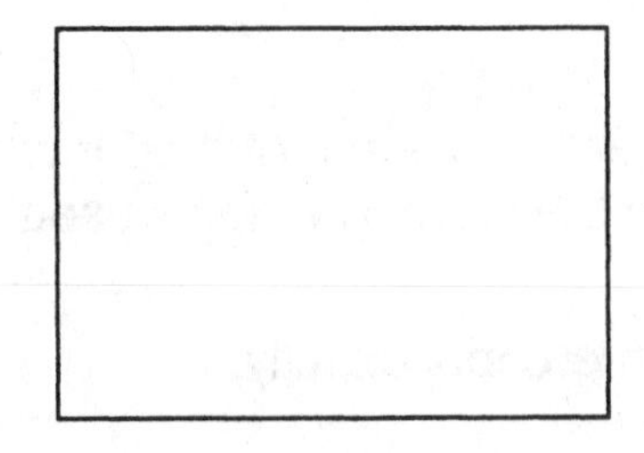

Answer:

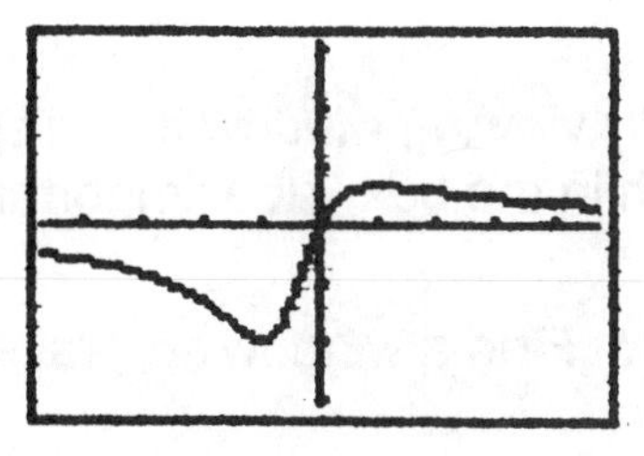

Using Graphical Methods to Solve Equations

There are 2 Basic Techniques to use the graphing capabilities of a calculator to solve an equation. ex. Solve $|x^2 - 2| = 4$ for x.

1. Rewrite the equation as $|x^2 - 2| - 4 = 0$
   The x-intercepts of the graph are the solutions to the equation.
   Enter $Y1 = |x^2 - 2| - 4$
   It is necessary to set the Xmax – Xmin so that this interval contains these intercepts.
   Set the window. Now any of the solve methods described for your calculator can be used.
2. Consider the equation as 2 functions. $Y1 = |x^2 - 2|$, $Y2 = 4$
   The intersections of the graphs are the points that satisfy both equations. The X coordinates of these points are the solutions of the equation.
   It is necessary to set the Xmax – Xmin and the Ymax –Ymin so that this window contains these intersections. Set the window. Now any of the solve methods described for your calculator can be used.

Comments:

If you can't find the required intersections on the screen method 1. may be better. You only need to look at 1 dimension, the x-axis, not in 2 dimensions, the viewing screen.

Method 1. is a special case of Method 2. where the second equation is $y = 0$.

Method 2. displays graphs of the 2 sides of the equation. The graph in Method 1. displays a new graph, the difference of the 2 functions.

Problems:

Find a friendly viewing window that displays a Complete Graph of each of the functions. Sketch the graph in the box, label important points, and state the viewing window used.

1. $y = x - 5$ Find a window so Trace will display both intercepts exactly.

Xmin =
Xmax =
Ymin =
Ymax =

2. $y = \frac{x^2-14x+13}{x-13}$ Find a window showing the discontinuity.

Xmin =
Xmax =
Ymin =
Ymax =

3. $y = \sqrt{121 - x^2}$ Find a Friendly Window for a True Shape Graph that displays the endpoints of this semicircle.

Xmin =
Xmax =
Ymin =
Ymax =

4. $y = 500(1.05)^x$ Find a Friendly Window for this model of $500 at 5% interest so that TRACE shows the values at $\frac{1}{2}$ year intervals for 40 years.

Xmin =
Xmax =
Ymin =
Ymax =

Answers:

Windows given by [Xmin, Xmax](Scl) by [Ymin, Ymax](Scl)

1. [-2.7, 6.7] by [-6.1, 0.1]

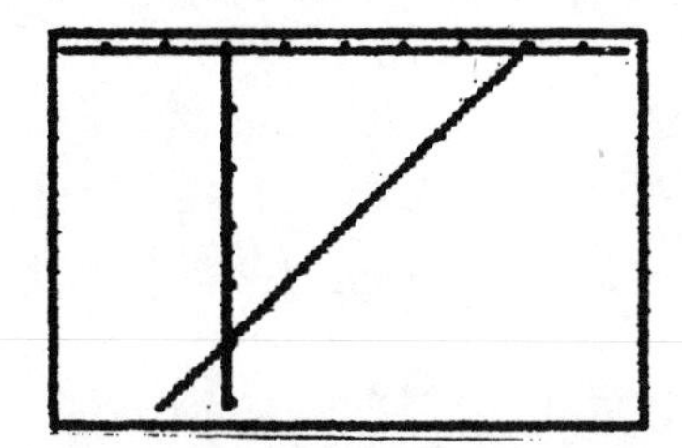

2. [8.3, 17.7] by [7.3, 16.7]

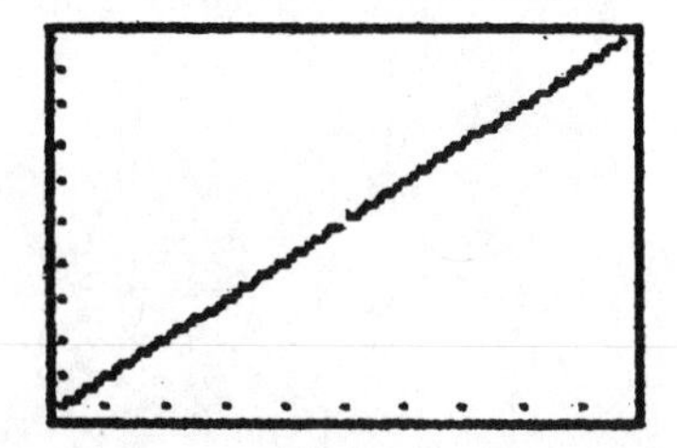

3. [-11.75, 11.75](5) by [-3.75,11.75]

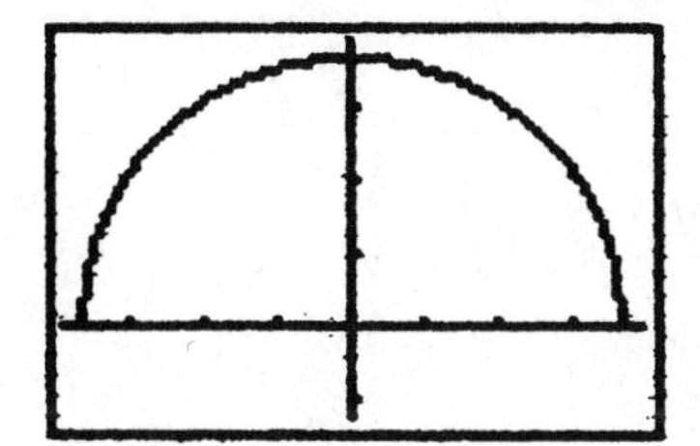

4. [0, 47](5) by [200, 5000](100)

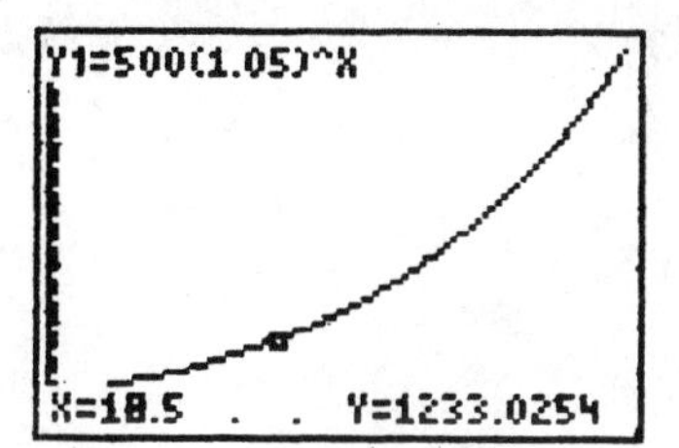

Chapter 2 - Calculator Notes and Problems

Basics

Scientific Nation:

A number written as a product of a number between 1 and 10 and a power of 10 is written in Scientific Notation.

ex. $123,456$ is written 1.23456×10^5

When calculators display a number in Scientific Notation, it is in a different form.

For example, 1.234×10^5, will be displayed $1.234E5$

To write a number in Scientific Notation from the keyboard, press the EE (E, EEX, EXP or 10^x) key for the exponent.

To enter $12,340,000,000$ press **1.234** [EE] **10**

To enter 0.002345 press **2.345** [EE] **− 3**

At some predetermined point, all calculators will change a very large or a very small number entered from the keyboard to Scientific notation.

ex. **.0123** × **.000000123** enter displays **1.5129 E − 9**

However calculations of numbers entered in Scientific Notation are displayed in Normal decimal notation unless the result is very large or very small.

ex. Enter the numbers 2.345×10^{-3} and 1.357×10^5 in Scientific Notation and multiply them.

Press **2.345** [EE] **−3** × **1.357** [EE] **5** enter

The display reads **318.2165**

To get all answers displayed in Scientic Notation , the Display must be changed to Scientific in the MODE (or SET UP) menu. (See the directions for your calculator). Some calculators allow for Floating Point when in Scientific Mode, others require that Fix is specified.

ex. Set the display from **Normal** to **Sci** in the **Mode** menu.

Set **Fix** at **3**.

Press **2.345** [EE] **−3** × **1.357** [EE] **5** enter

The display reads **3.182E2**

Problems:

Set the Display to Scientific Notation,

Enter the numbers in Scientific Notation, then

Evaluate

1. $(3.24 \times 10^4)^3$
2. $(12,700) \times (2,370) \div (.0493)$

Answers:

1. $\underline{3.40122E13}$
2. $\underline{6.10527E8}$

Order

Sometimes Calculators use inequalities to represent intervals on the X-axis See the directions for your calculator on entering split functions.

Negative Numbers and Subtraction

The operation of making a number negative is not the same as the operation of subtraction. There are different symbols on graphing calculators for the different operations. Use (-) to enter a negative number. Use – to indicate subtraction. The subtraction symbol can never be used at the beginning of a line.

ex. Subtract 5 from negative 3.

Press (-) $3 - 5$enter, the result is $\underline{-8}$

Checking Answers or "Is the point on the graph?"

The solutions (answers) to Algebraic problems are coordinates of the graphs that represent the problems. To find out if a point is on a graph, TRACE can be used on a suitable window or a TABLE of the values of the function may contain the point. However, it is faster and easier to check directly on the home screen.

ex. Is the point $(1,-2)$ on the graph of $3x - y = 5$?
Store 1 in x, store -2 in y, then enter $3x - y$ on the home screen, enter, the result is $\underline{5}$
The point is on the graph.

ex. Is $x = 1$ the solution of $3x - 15 = 12$?
Store 1 in x, then enter $3x - 15$ on the home screen, enter, the result is $\underline{-12}$
So $x = 1$ is not the solution.

Is $x = 9$ the solution of $3x - 15 = 12$?
Store 1 in x, then enter $3x - 15$ on the home screen, enter, the result is $\underline{12}$
So $x = 9$ is the solution.

ex. Is the point $(\sqrt{2}, -.757)$ on the graph of $3x - y = 5$
Store $\sqrt{2}$ in x, store $-.757$ in y, then enter $3x - y$ on the home screen, enter, the result is $\underline{4.99640687...}$
The point is not on the graph.

ex. Is the point $(\sqrt{2}, 3\sqrt{2} - 5)$ on the graph of $3x - y = 5$

Store $\sqrt{2}$ in x, store $3\sqrt{2} - 5$ in y, then enter $3x - y$ on the home screen, enter, the result is $\underline{5}$

The point is on the graph.

ex. Solve the equation $5x^2 - 3x - 6 = 0$ for the exact answers and check them.

Use the discriminent. $a = 5, b = -3, c = -6$

$(-3)^2 - 4 \times 5 \times -6 = 129$

Use the quadratic Formula

$x = \frac{+3+\sqrt{129}}{10}$ $\quad x = \frac{+3-\sqrt{129}}{10}$

Check the first answer

Store $\frac{+3+\sqrt{129}}{10}$ in x, enter $5x^2 - 3x - 6$ on the home screen, enter, the result is $\underline{0}$

Now check the second answer.

Note the purpose of each parenthesis in the last example.

store $((3 + \sqrt{\ }(129)) \div 10)$ in x

The outside set contains the expression to store.

The next set contains the numerator of the fraction.

The inside set contains the number for the square root.

Problem:

Solve and check the following equations. If the answer has an exact and approximate form, check both. Use 3 place accuracy.

1. $x + 5 = 2x - 7$
2. $2x^2 + 12x + 5 = 0$
3. $x^2 + 3x + 4 = 5$

Answers

1. $x = 12$
2. $x = -3 \pm \sqrt{\frac{13}{2}} \approx -.450, -5.550$
3. $x = \frac{-3\pm\sqrt{13}}{2} \approx .303, -3.303$

Absolute Values

The absolute value is written as *abs* on a graphing calculator.

$|-4| = abs(-4) = 4$

$|2^2| = abs(2^2) = 4$

When evaluating numeric expressions involving an absolute value

$4 - |3 - 9|$ would be entered $4 - abs(3 - 9)$

Graph $y = |x|$ on the decimal window.

You can see that $|c| = c \quad if\ c \geqq 0$

$|-c| = -c \quad if\ c < 0$

Trace to confirm the values.

Absolute value equations and inequalities can be solved

graphically on a graphing calculator.

ex. Solve $|x+5| < 3$

Enter Y1 $= abs(x+5)$ and Y2 $= 3$.

Graph on Window. $[-4.7, 4.7]$ by $[-1.1, 5.1]$

Trace to the intersection $(-2, 3)$

The line $y = x + 5$ is below the line $y = 3$ for $x < -2$.

The answer is $x < -2$.

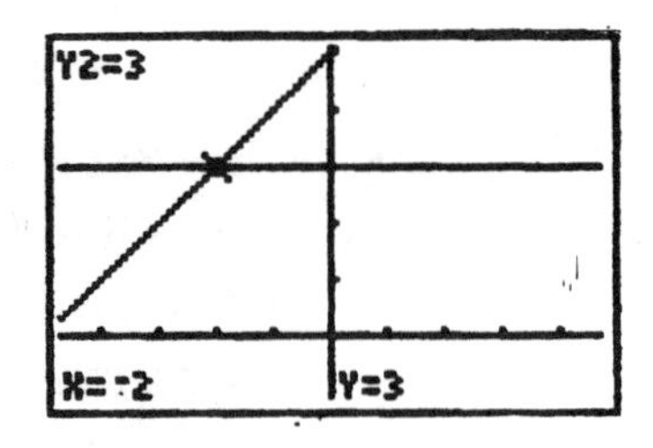

Problems

1. $|2x+3| = 9$
2. $|2x+3| = 4x - 1$
3. $|x^2 - 4x - 1| = 4$
4. $|x+4| \le 2$
5. $|x-6| > 2$

Answers

1. Graph $|2x+3| - 9$ on the decimal window. Zoom out by a factor of 2, This is still a Friendly window. TRACE $x = -6$ and $x = 3$

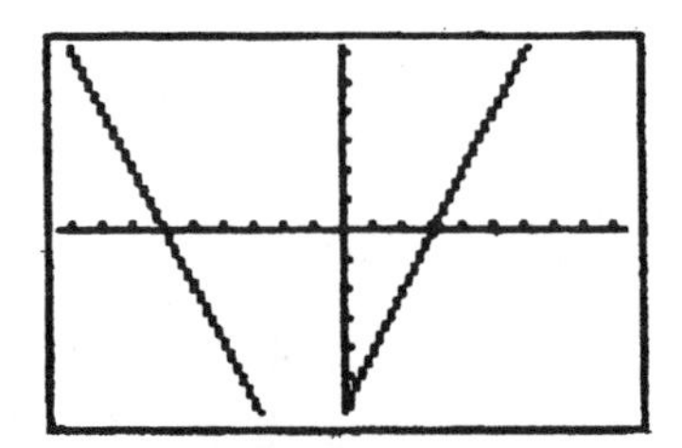

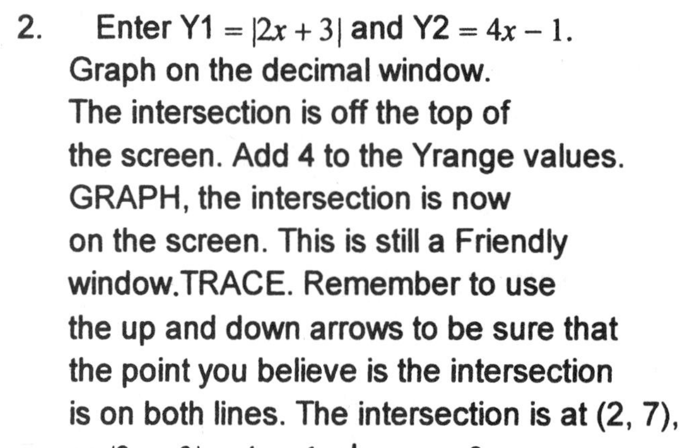

2. Enter Y1 $= |2x+3|$ and Y2 $= 4x - 1$. Graph on the decimal window. The intersection is off the top of the screen. Add 4 to the Yrange values. GRAPH, the intersection is now on the screen. This is still a Friendly window.TRACE. Remember to use the up and down arrows to be sure that the point you believe is the intersection is on both lines. The intersection is at (2, 7), so $|2x+3| = 4x - 1$ when $x = 2$.

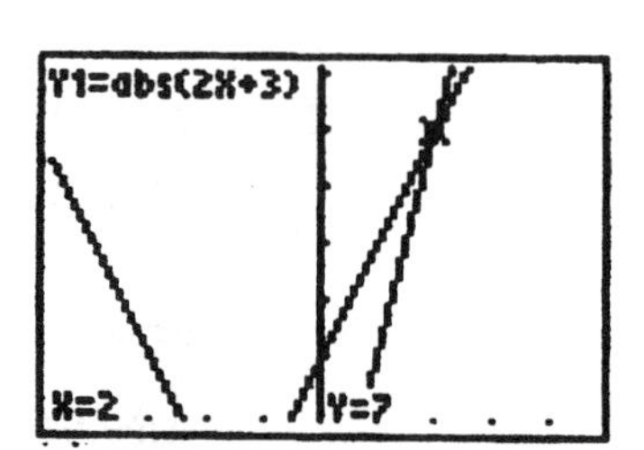

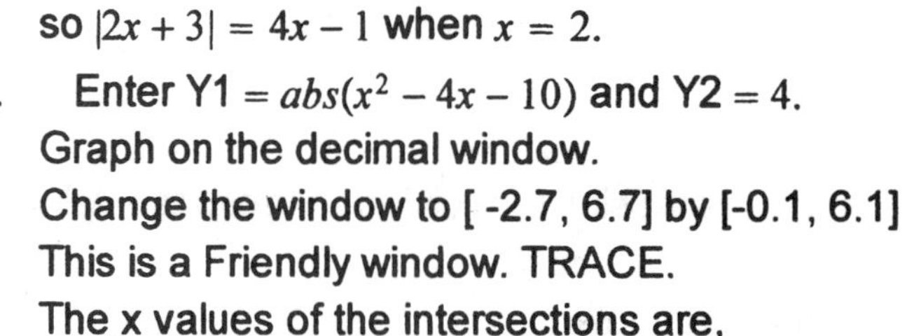

3. Enter Y1 $= abs(x^2 - 4x - 10)$ and Y2 $= 4$. Graph on the decimal window. Change the window to [-2.7, 6.7] by [-0.1, 6.1] This is a Friendly window. TRACE. The x values of the intersections are,
$$x = -1, 1, 3, 5.$$

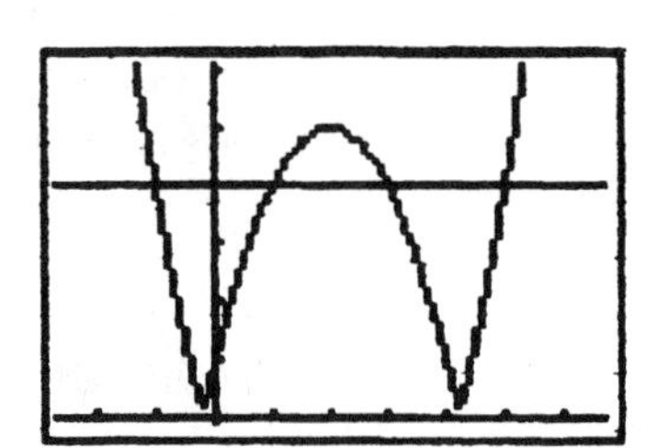

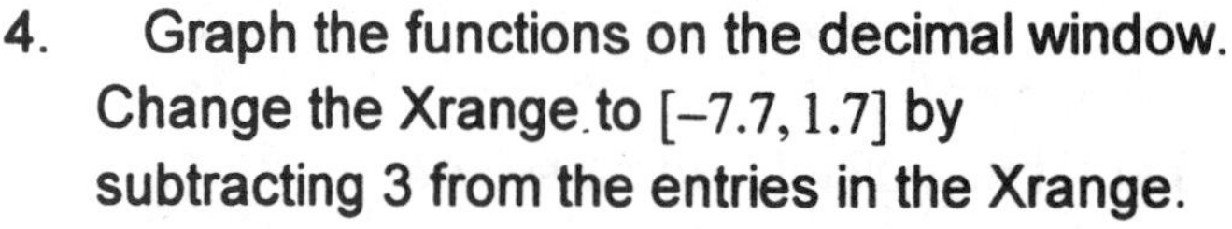

4. Graph the functions on the decimal window. Change the Xrange to $[-7.7, 1.7]$ by subtracting 3 from the entries in the Xrange.

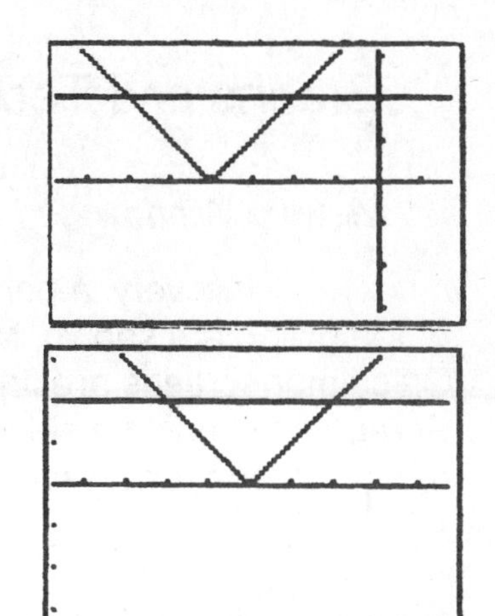

This is a Friendly Window. TRACE
The intersections are at $(-6, 2)$ and $(-2, 2)$
$|x+4| \le 2$ for x between -6 and -2, $-6 \le x \le -2$

5. Graph the functions on the decimal window.
Change the Xrange to $[1.3, 10.7]$ by adding 6 to the entries in the Xrange.
This is a Friendly Window. TRACE
The intersections are at $(4, 2)$ and $(8, 2)$
$|x-6| > 2$ for $x < 4$ or $x > 8$.

Graphing a Circle:

ex. Graph the circle $(x+1)^2 + (y-3)^2 = 9$
From the equation, the center of the circle is at the point $(-1, 3)$ and the radius is 3.The diameter of the circle is 6, so both Xrange and Yrange must be at least 6 units long. with the center at the point $(-1, 3)$.
Solve the equation for y, then enter

$$Y1 = 3 + \{-1, 1\}\sqrt{9-(x+1)^2}$$

Graph on the decimal window.
Change the Yrange to $[-.1, 6.1]$
The graph is a circle.

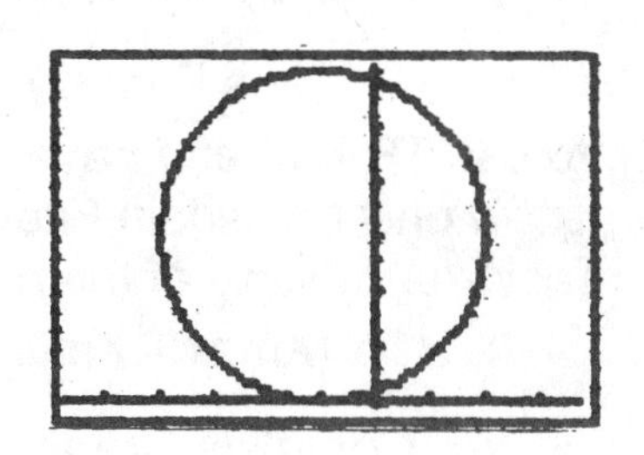

Graphs and Technology

Viewing Window

It is very important to get a good viewing window so that the important properties of the graph are shown. Many functions have the set of Real Numbers for their domain. Calculator screens are finite in size, so they can not possibly show all of the function. So, a Screen size must be selected to view the interesting parts of the graph. Sometimes 2 or more screens are needed to see both the local and global properties of the graph.

True Shape

Graphs are their true shape if the X pixel size is the same as the Y pixel size. Pixel size = (Xmax – Xmin) ÷ (number of pixels on the width of the screen) = (Ymax – Ymin) ÷ (number of pixels on the height of the screen). On the Decimal and Integer Windows graphs are their true shape. See the Basic Calculator Topics chapter for a table of Screen sizes for different calculators.

Friendly Window

On a Friendly Window each pixel has a "nice" size, 0.1, 0.2, 0.5, 1 etc. It is easy to use TRACE and have a good idea what the coordinates of a point are. "Holes" in graphs show only on certain Friendly Windows. Friendly Windows occur when Xmax – Xmin is a factor or multiple of the number of pixels on the length of the calculator screen, or

(Xmax – Xmin) = k (number of pixels).Decimal windows are always friendly.

Complete Graph

A Complete Graph shows all of the important properties of the function: intercepts, local maximum, local minimum, vertical asymptotes and end behavior. Often more than one window is needed to show all of these properties.

Numerical Solutions

Each calculator has built in Algorithms to find Numerical Solutions, really approximations, to certain problems. They all have a numeric root (intercept) program. The program will ask you for a guess (seed) so that it has a starting point for the numerical method. The guess should be near the required root. For more than one root, each guess should be nearer the root wanted than to any other root. Graph the function on a Friendly Window to find guesses for the program.
Directions are included for each calculator.

Graphic Solutions

Using TRACE and ZOOM it is possible to get solutions to Algebraic problems. These solutions will be more accurate if the function is graphed on a Friendly Window. Details of the methods are given in each set of calculator directions. Some Graph Solve routines do not need a guess.

Graphing Convention

Chapter 2 – Calculator Notes and Problems

1. Complete graphs are required unless a Viewing Window is specified or the problem indicates that a partial graph is acceptable.

2. If the directions say "find the graph", "obtain the graph" or "graph the equation", graph it on your calculator. You do not have to copy it to paper.

3. If the directions say "sketch the graph" it must be drawn on paper, indicating the Xmax, Xmin, Ymax and Ymin. Copy the calculator graph. If the graph is not clear on the calculator, graph with pencil and paper.

Problems

1. Find a Friendly Window that shows a Complete Graph of
   $y = 0.25x^4 - 1.5x^3 - 15x^2 + 19x + 100$
   Hint: the graph has 4 distinct roots and enters in the upper left.
2. Enter the following data in the STAT menu.
   x is the number of minutes of Advertising on TV
   y is the profit in sales for the next week in ten thousand dollars

x	1	2	3	4	5	6
y	12.0	10.8	15.0	22.0	22.6	28.9

   Find a Linear Regression that approximates the data.
   Draw the Scatter Plot and the Regression Line.

Answers

1. *There is more than one correct answer.*
   Enter the equation in the Y= menu.
   Graph on the decimal window.
   There are 2 roots shown.
   Zoom out to see all 4 roots.
   The X interval is to big, so TRACE
   to get the approximate values for
   the smallest and largest roots.
   They are near $x = -6$ and $x = 11$ All the roots will fit
   on an interval of length $23.5 = .5(47)$
   Pick Xmin = -9 and Xmax = 14.5 (Other values are O.K.)
   Set these values in the viewing window.
   Use the AutoScale (ZoomFit on a TI.)
   The graph fits in the window, but the local behavior
   is not clear.
   Change Ymin = -500, Ymax = 750, Yscl = 100
   Graph. Here is a clear Complete Graph on a Friendly
   Window.

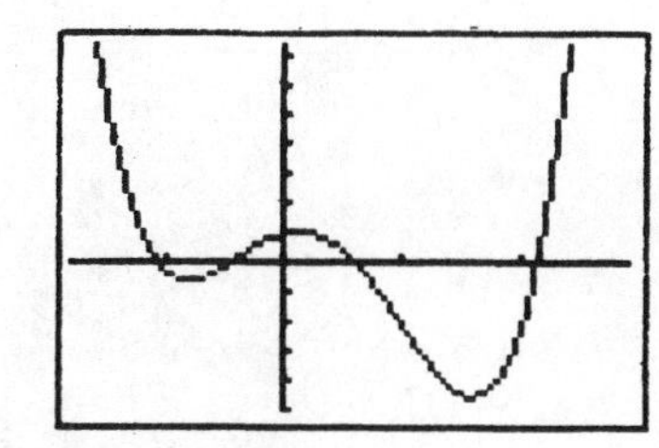

Use TRACE, each x pixel is 0.25

2. *The specific keystrokes for your calculator are in the calculator directions.*

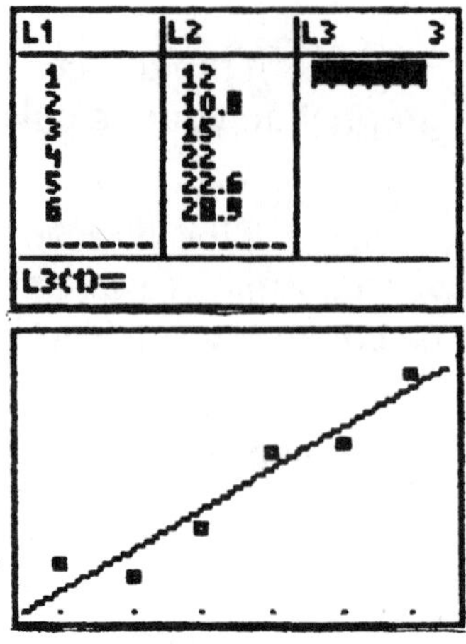

Enter the data in the Statistic Lists using EDIT.
Calculate the Linear Regression equation.

$y = 3.9825x + 5.275$

Set the Stat Plot to Scatter, x in List 1, y in List 2.
Graph the Scatter Plot.
Turn off or delete the equations in the Y= menu.
Paste the Linear Regression equation into the Y= menu.
Graph the Regression Line on the Scatter Plot.

Remember to turn off the Plot Menu when you are finished.

Highlight PLOT 1, then press ENTER

Functions and Graphs

What is a function?

A function is a RULE. Given an input x, the rule does something with x and gives back a number y, sometimes written as $f(x)$.

ex. Suppose the rule is cube x, take its negative, add 2 times x squared, add 5 times x then subtract 6.

The rule can be written as an Algebraic formula,

$$y = -x^3 + 2x^2 + 5x - 6$$

If this rule is entered into the Y= menu as Y1 on a TI calculator, function y values can be computed on the home screen.

$Y1(3) = 0, \quad Y1(5) = -56.$

You supply the number for x, the calculator uses the rule for y and displays the numeric value of y with the rule and the number you picked for x.

(unfortunately, functional notation only works on a TI)

Plot1 Plot2 Plot3
\Y1 = -X^3+2X²+5X-6
\Y2=
\Y3=
\Y4=
\Y5=
\Y6=

The formula is the Algebraic representation of the function.

The formula could be used in a TABLE. If you make a list of numbers to use for x, the calculator displays a list of numbers that have been computed from the rule.

For each x it computes the correct value of y that is in the TABLE.

A TABLE that had every number that could be used for x is impossible to write, so the TABLE only has some ot the possible inputs for the function.

X	Y1
0	-6
1	0
2	4
3	0
4	-18
5	-56
6	-120

X=0

A TABLE is the Numeric representation of the function.

The numbers in the TABLE can be thought of as a pair of numbers where the x is first number and the y is second number.

This pair is written (x, y).

This pair can be thought of as a point in the XY plane. If the points in the TABLE are graphed as points in the XY plane,and these points are connected with a smooth curved line, then this line is called the graph of the function (or Rule).

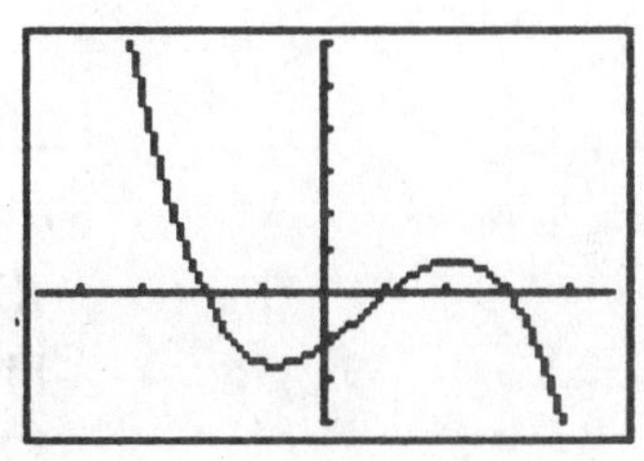

called the graph of the function (or Rule).

The Graph is called the Graphic representation of the function.

A function consists of a set of inputs called the domain, a rule that determines how to manipulate the number from the input, and a set of outputs from the rule called the range. There must be only one output for each input.

Function Machine

Your graphing calculator is a Function Machine. Different Rules are marked on the keys: x^2, $\sin(x)$, $\log(x)$, $\sqrt{x}$.

You pick the x, the calculator knows the rule, it displays the y.

For each x that is picked, only one y is shown.

The x is called the argument, the y is called the value of the function.

The set of $x's$ used as input is called the domain,

The set of $y's$ for output is called the range.

Rule of Three

The rule that was given for the function: cube x, take its negative, add 2 times x squared, add 5 times x then subtract 6, is given in words.

The Rule of 3 states that there are 3 ways to represent a function in mathematics: as a formula, a table or a graph. All represent the same function or rule.

Piecewise Defined Functions

A piecewise defined function is one whose rule includes more than one formula.

The way to enter a piecewise defined function in your calculator is included in the calculator directions.

ex. Graph the piecewise defined function

$f(x) = x + 1 \quad -2 \le x < 1$

$\qquad 1 + 2x \quad 1 \le x \le 5$

Then TRACE to find

$f(0)$, $f(-1)$, $f(1)$,

$f(2)$, $f(5)$, $f(6)$

$Y1 = (x + 1)(-2 \le x < 1)$

$Y2 = (1 + 2x)(1 \le x < 5)$

Graph on the Decimal Window.

Change the Window $[-2.9, 6.5][-2, 15]$

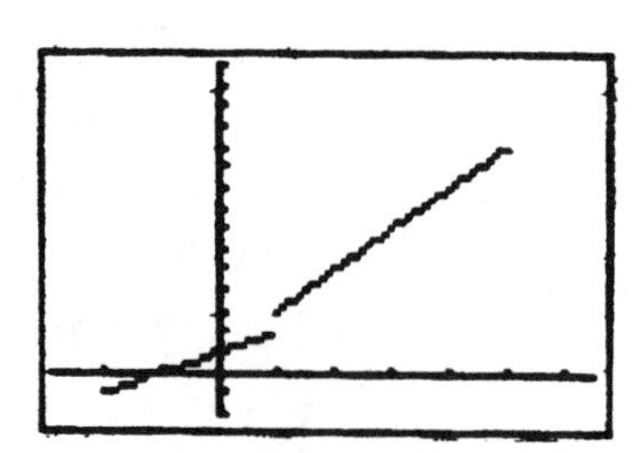

This is a Friendly Window. TRACE
Use the up and down arrow keys to move
between the 2 graphs.
Read $x = -1,\quad y = 0$
$x = 0,\quad y = 1$
$x = 1,\quad y = 3$
$x = 2,\quad y = 5$
$x = 5,\quad y = 11$
$x = 6,\quad y = ?$

The last entry may have $y = 0$, where it is not defined. This might not be true.

Notice that the inequality shows which piece of the equation to use for the function rule.

Problems

Graph the piecewise defined functions on a Friendly Window, then TRACE to find:
$f(0),\ f(1),\ f(2),\ f(3)$

1. $f(x) = 1 - 3x \qquad x < 2$
   $\quad\ = 2x + 3 \qquad x \geq 2$

2. $f(x) = x \qquad x \leq 1$
   $\quad\ = x^2 \qquad x > 1$

Answers:

1. Window: $[-4.7, 4.7]$ by $[-5.3, 13.3](2)$
   $f(0) = 1$
   $f(1) = -2$
   $f(2) = 7$
   $f(3) = 9$

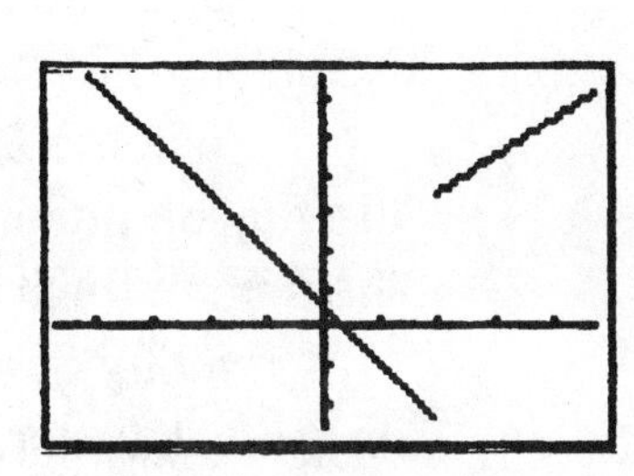

2. Window: $[-4.7, 4.7]$ by $[-1.1, 5.1]$
   $f(0) = 0$
   $f(1) = 1$
   $f(2) = 4$
   $f(3) = 9$

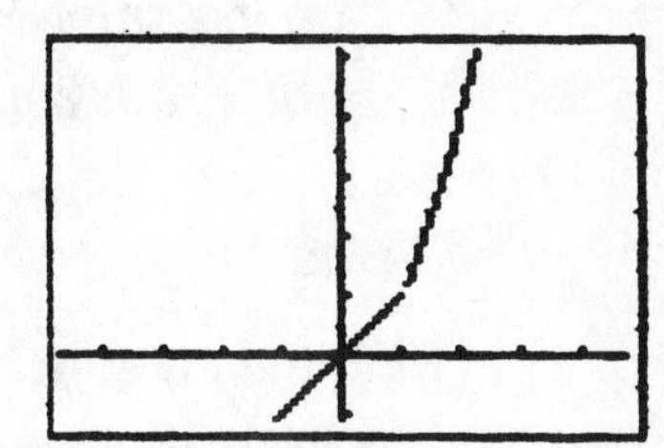

Parametric Graphs

When the graph of a curve fails the Vertical Line test, it can not be graphed as a function. The x and y coordinates can both be written as functions of a third variable t. The equations $x = f(t)$ and $y = g(t)$ are called parametric equations. The curve can be graphed by changing from Funtion to Parametric MODE.

TRACE works in Parametric Mode, T, X and Y are all displayed.

The parametric equation values can be displayed in a TABLE. The first column is T, the independent variable, the next 2 columns display the values of X and Y.

Operations on Functions

When some functions are entered into a calculator, care must be taken to include the parenthesis that tell the calculator how to combine the numbers.

Products: $(x^2+2)(x-3 \div x)$

Quotients: $(x^2+2) \div (x-5)$

Powers: (x^2+1)^$(x-2)$

Composition: $\sqrt{\ }(x^2+2)$ is $f \circ g$,

where $f(x) = \sqrt{x}$, $g(x) = x^2+2$

When in doubt **Use Parenthesis**.

Inverse Functions

Given the function $y = \sqrt{x-2}$ with domain $x \geq 2$, range $y \geq 0$.

This function can be solved for x.

$$y = \sqrt{x-2}$$
$$y^2 = x-2$$
$$x = y^2+2$$

Interchange x and y. $\quad y = x^2+2$

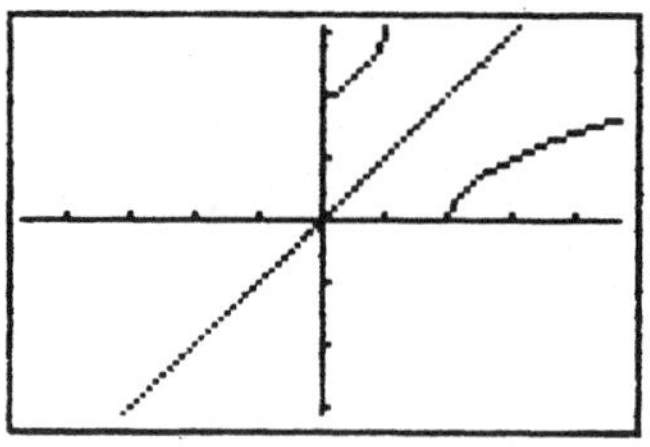

This last equation is the inverse function.

It has domain $x \geq 0$, and range $y \geq 2$.

Graph $y = \sqrt{x-2}$, $y = x^2+2$, $y = x$ on the same screen:

(use the Decimal window)

The funtion and its inverse are symmetric about the line $y = x$.

The screen must give the True Shape of the graphs or the symmetry does not show.

On the calculator, functions and their inverse functions are on the same key, the inverse is the 2nd function.

For x^2 the 2nd function is $\sqrt{\ }$.

Problems:

Find the inverse function for the following functions. Graph the function, the inverse and the line $y = x$ on a square window.

1. $y = \frac{2}{x}$
2. $y = 3x^3 - 1$
3. $y = x^3 + x + 1$ Hint: Use parametric equations.

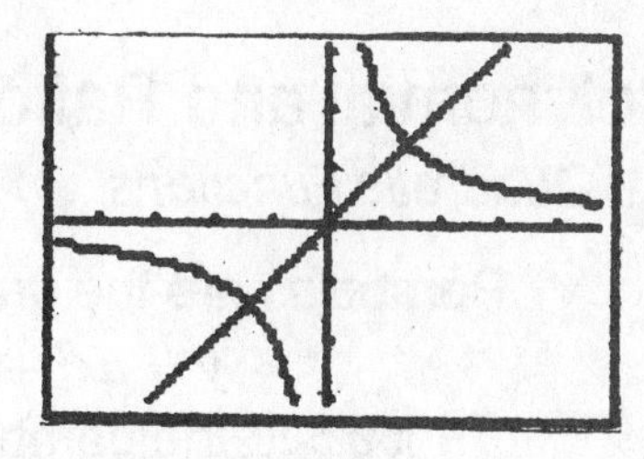

Answers:

1. Window: $[0, 9.4]$ by $[0, 6.2]$
   $y = \frac{2}{x}$ is it's own inverse.
   Why is this?
   The domain is all real x
   The range is all real y.

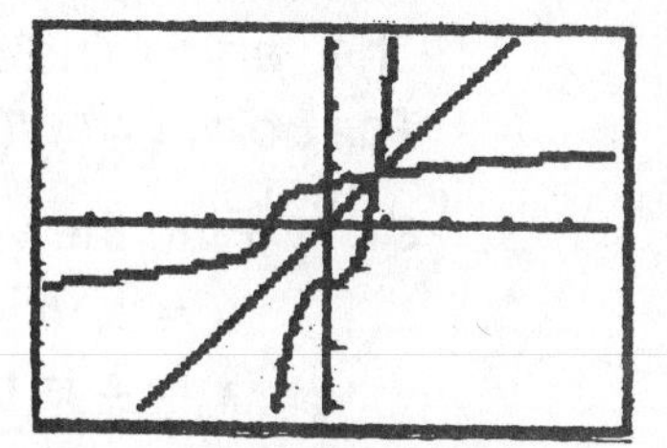

2. Window: Decimal
   $y = \sqrt[3]{\frac{x+1}{3}}$ is the inverse.
   The domain is all real x
   The range is all real y.

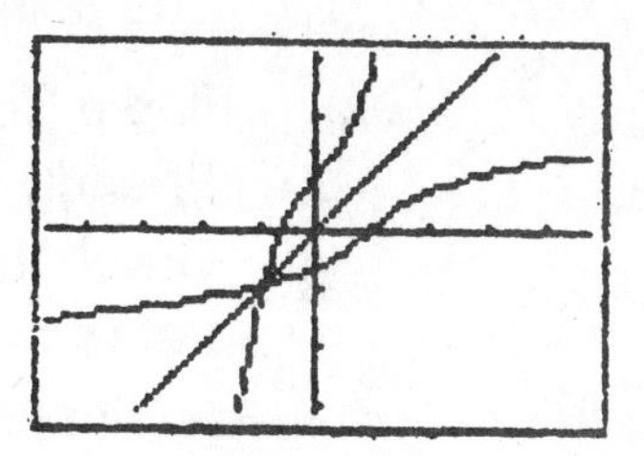

3. There is no algebraic way to solve for the inverse function.
   For the function use: $x = t,\ y = t^3 + t + 1$
   For the inverse use: $x = t^3 + t + 1,\ y = t$
   The domain is all real x
   The range is all real y.

Polynomial and Rational Functions

Quadratic Functions

Parabolas are the graphs of quadradic functions of the form

$$y = ax^2 + bx + c$$

If $a > 0$, the parabola opens up, the vertex is at the bottom.

If $a < 0$, the parabola opens down, the vertex is at the top.

The vertical line through the vertex is the axis of symmetry.

The point $(\frac{-b}{2a}, f(\frac{-b}{2a}))$ is the vertex

ex. Find the vertex of $y = x^2 - 3x - 4$

$a = 1,\ b = -3,\ c = -4$

$x = \frac{3}{2}$ is the equation of the line of symmetry

Store $\frac{3}{2}$ in x, evaluate y at $x = \frac{3}{2}$, $y = -6.25 = \frac{-25}{4}$

The point $(\frac{3}{2}, \frac{-25}{4})$ is the vertex of the parabola.

ex. Find a Complete Graph of $y = x^2 - 3x - 4$ on a Friendly Window.

Graph the equation on the decimal Window.

Let (Ymax - Ymin) = 2(6.2),

then translate 2 units down.

Ymin = -8.1, Ymax = 4.1 TRACE

The x intercepts are $x = -1, x = 4$.

The y intercept is $y = -4$

The vertex is (1.5, -6.25)

The minimum value of the function is -6.25

located at $x = 1.5$

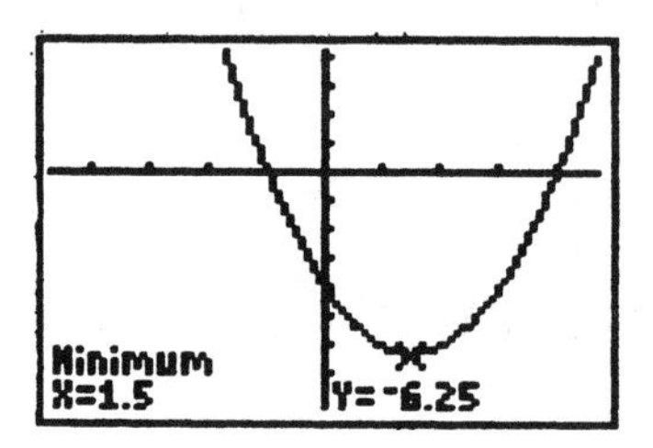

Problems:

Find a Complete Graph of the quadratic equations on a Friendly Window.

Find the intercepts, axis of symmetry, vertex, and the maximum or minimum value for y.

Find exact values if possible, otherwise approximate.

1. $y = -x^2 + 2x + 48$
2. $y = 2x^2 + 12x + 5$

Answers

1. A Friendly Window is

[-9.4, 9.4](2) by [-40, 53](10)

The intercepts are $x = -6$ and $x = 8$

The vertex is $(1, 49)$

Axis of Symmetry, $x = 1$

Maximum value $y = 49$

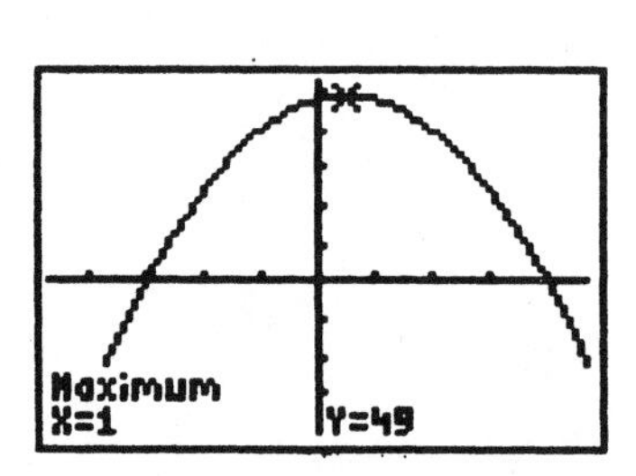

Maximum value $y = 49$

2. A Friendly Window is
[-8.7,0.7] by [-20, 42](10)
The intercepts are approximately
$x = -5.549$ and $x = -.451$
The vertex is $(-3,-13)$
Axis of Symmetry, $x = -3$
Minimum value $y = -13$

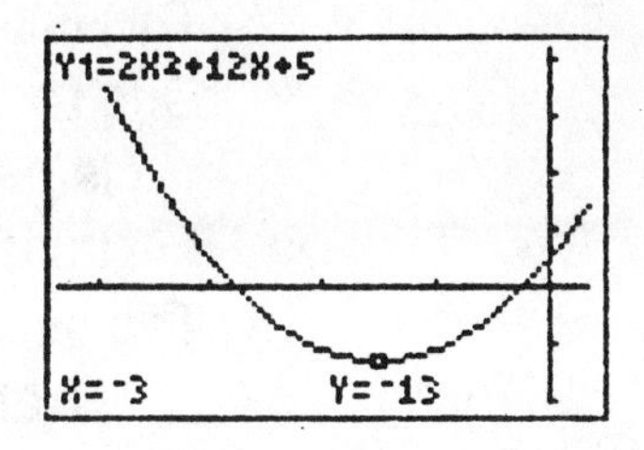

Polynomials

Remainder Theorem: To find the remainder when $f(x)$ is divided by $x - c$, find $f(c)$.
The remainder is the same as the value of y at $x = c$.

ex. Find the remainder when $f(x) = 3x^4 - 8x^2 + 11x + 1$ is divided by $x + 2$. By the Remainder Th. compute $f(-2)$.
Store -2 in x. Evaluate $f(x)$ at $x = -2$. $f(-2) = -5$.

A polynomial of degree n has exactly n roots counting multiple roots and complex roots.

Rational Root Test:
If a rational number $\frac{r}{s}$ (in lowest terms) is a root of the polynomial

$$a_n x^n + \ldots + a_1 x + a_0$$

where the coefficients $a_n, \ldots, a_1, a_0$ are integers with $a_n \neq 0, a_0 \neq 0$, then
r is a factor of the constant term a_0
s is a factor of the leading coefficient a_n

When the coefficients of a polynomial are Integers, the irrational roots and the complex roots occur in pairs.

ex. Find the possible Rational Roots $\frac{r}{s}$ of
$y = 6x^3 - 19x^2 + 16x - 4$
The numerator r is a factor of 4
The denominator s is a factor of 6
Make a table

$s\backslash r$	± 1	± 2	± 4
1	± 1	± 2	± 4
2	$\pm\frac{1}{2}$	$\pm\frac{2}{2}$	$\pm\frac{4}{2}$
3	$\pm\frac{1}{3}$	$\pm\frac{2}{3}$	$\pm\frac{4}{3}$
6	$\pm\frac{1}{6}$	$\pm\frac{2}{6}$	$\pm\frac{4}{6}$

Reducing and eliminating repetitions, the possibilities are

$x = 1, 2, 3, \frac{1}{2}, \frac{1}{3}, \frac{2}{3}, \frac{4}{3}, \frac{1}{6}$

Graph $f(x)$ on the Decimal Window.
Change the Xrange to [0, 4.7]
This is a Friendly Window, TRACE.
One root is $x = 0.5$, the second root is between $x = .6$ and $x = .7$, the third root is $x = 2$.
The second root must be rational, (irrational root occur in pairs)
The only possibility is $x = \frac{2}{3} = .66\overline{6}$
Check these values in a TABLE.

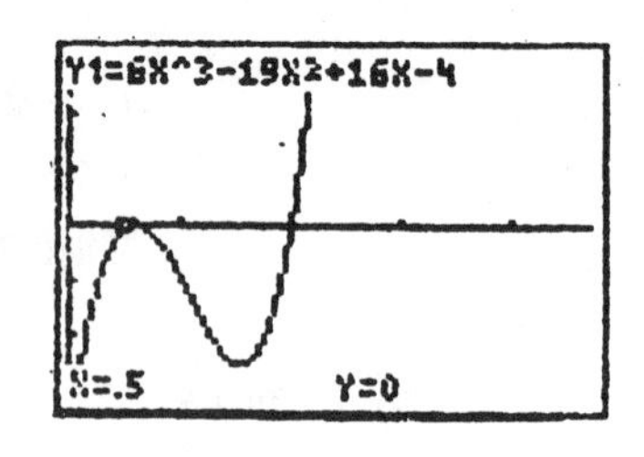

To Find the Roots of a Polynomial from a Graph

1. Make a list of the possible Rational Roots.
2. Pick Xmax - Xmin so that this interval contains all of the possible Rational Roots. Make this a Friendly interval.
3. Regraph with Xmax - Xmin = k(96) where the visible roots are on this interval.
4. TRACE to see which of the possible roots are on the graph.
5. Check these values on the Home Screen or in a TABLE.
   To make this a Complete Graph, use the Autoscale feature.

Polynomial Root Finders

The TI 85, 86, 89, 92, Sharp 9600, HP 38
Casio 9850, degree ≤ 3, fx 2, degree ≤ 30
all have Polynomial Root Finders.
Enter the coefficiants of the polynomial, press SOLVE and the rational, irrational and complex roots are displayed as decimal approximations.

ex. $f(x) = x^5 + 2x^4 - 6x^3 + x^2 - 5x + 1$
has the roots $x = 1.7964, 0.1191, -3.7636$
and the complex pair $x = -0.116 \pm 0.854i$

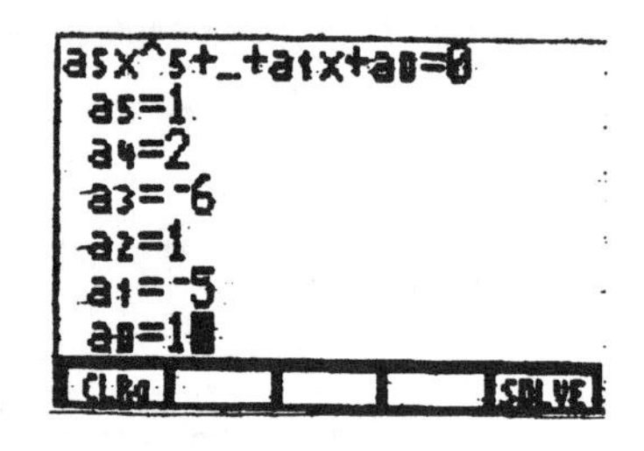

Numeric Solver and Graph Solver

These Solvers only find the Real roots of the function. Some of them need a "*close guess*" to compute each root.

Problems

Find all the Real roots of the polynomials. If you have a

Polynomial Solver, find all of the zeros, real and complex.

1. $y = x^5 - 2x^4 + x^3 - x^2 + 2x - 1$
2. $y = x^5 - x^4 + x^3 - 13x^2 + x + 6$
3. $y = -2x^4 + 13x^3 - 21x^2 + 2x + 8$
4. $y = x^4 - 13x^2 - 12x$
5. $y = x^5 - x^4 + 3x^2 + 3x$

Answers

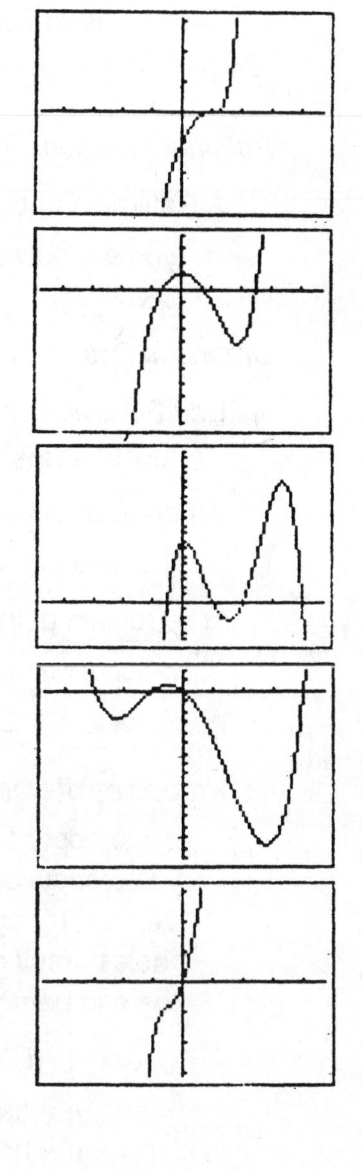

1. Decimal Window
   $x = 1, 1, 1, -.5 \pm .866i$

2. [-4.7, 4.7] by [-50, 20](5)
   $x = 2.4828, 0.7369, -0.6152,$
   $x = -0.8023 \pm 2.1649i$

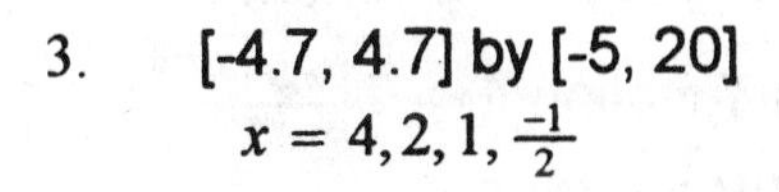

3. [-4.7, 4.7] by [-5, 20]
   $x = 4, 2, 1, \frac{-1}{2}$

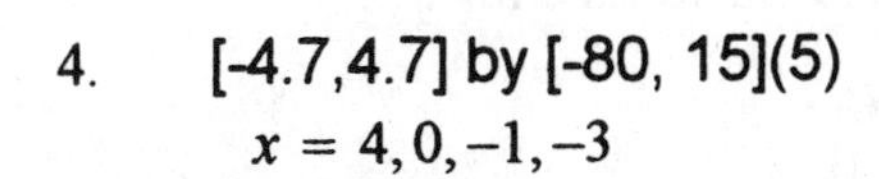

4. [-4.7,4.7] by [-80, 15](5)
   $x = 4, 0, -1, -3$

5. Decimal Window
   $x = 0$
   $x = 1.308 \pm 1.269i, \ -.808 \pm .501i$
   Note that technology is necessary to find these 4 complex roots.

Quadratic Regression

Problem

Enter the following data in the STAT menu.

x is the number of minutes of Advertising on TV

y is the profit in sales for the next week in ten thousand dollars

x	1	2	3	4	5	6
y	12.0	10.8	15.0	22.0	22.6	28.9

Find a Quadratic Regression that approximates the data.
Draw the Scatter Plot and the Regression Curve.

Answer

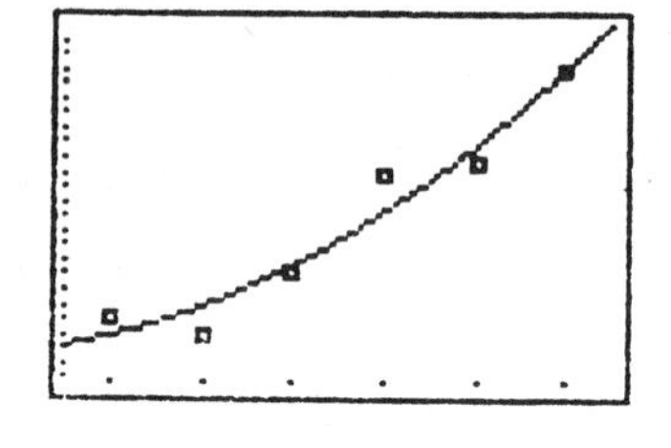

The regression equation is
$y = .8379x^2 - 1.254x + 11.5591$

Rational Functions

A Rational Function is of the form $f(x) = \frac{p(x)}{q(x)}$ where $p(x)$ and $q(x)$ are polynomials.

$p(x) = a_n x^n + \ldots + a_0 \qquad q(x) = b_m x^m + \ldots + b_0$

Graphs of Rational Functions

End Behavior: When $|x|$ is very large, the graph of a polynomial function closely resembles the graph of its highest degree term.

When $|x|$ is very large, the graph of the rational function $f(x) \approx \frac{a_n x^n}{b_m x^m}$ the ratio of the coefficients of the highest degree terms.

The Domain of the rational function $f(x) = \frac{p(x)}{q(x)}$ is the set of all rational numbers that are **not** roots of the denominator $q(x)$.

Big-Little Rule: $\frac{1}{BIG} = LITTLE, \quad \frac{1}{LITTLE} = BIG$

Properties of Rational Functions:

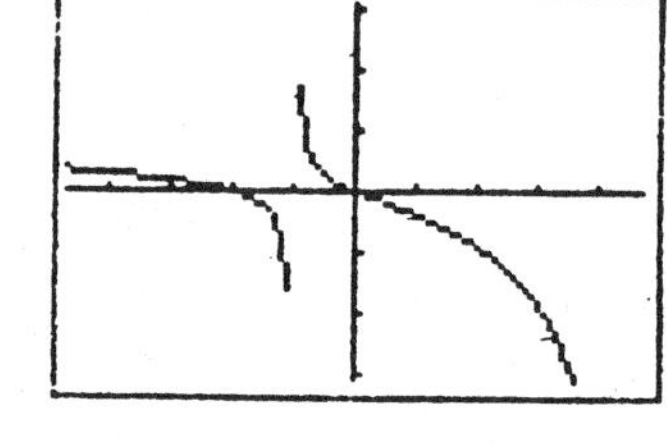

x intercepts occur when $p(x) = 0$ and $q(x) \neq 0$
y intercepts occur when $x = 0$
vertical asymptotes occur when $q(x) = 0$ and $p(x) \neq 0$
"holes" may occur when $p(x) = 0$ and $q(x) = 0$
the end behavior is $y = \frac{a_n x^n}{b_m x^m}$

ex. $f(x) = \frac{x^2+2x}{(x^2-4x-5)}$

$f(x)$ has roots at $x = 0, x = -2$
$f(x)$ has vertical asymptotes at $x = -1, x = 5$
End behavior as $|x|$ gets large $f(x) \approx \frac{x^2}{x^2} = 1$
so $y = 1$ is the horizontal asymptote
The decimal Window shows both roots and the asymptote $x = -1$
The Standard Window shows the asymptotes

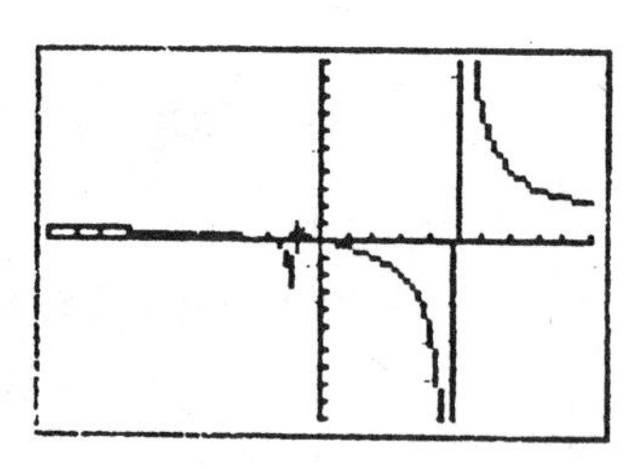

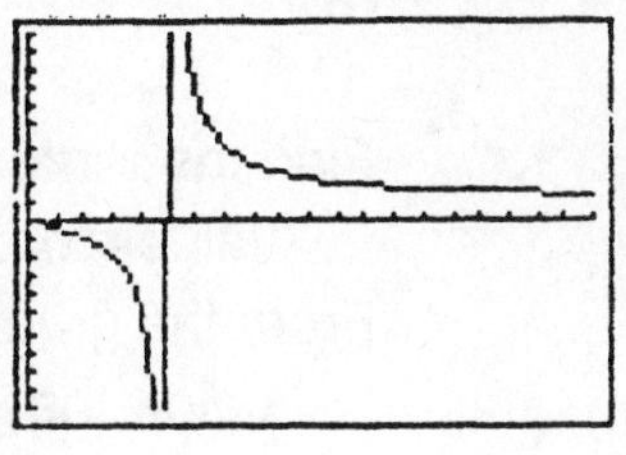

$x = 5, y = 1$ to the left
Change Xrange to [0, 20] shows $y = 1$ to the right.

Helpful Hints:

A Friendly Window will show less technology errors than a window that is not Friendly.

If the graph is not clear because of the extra technology error lines, try DOT rather than CONNECTED MODE.

Problems

1. $f(x) = \frac{3-2x}{x-1}$
2. $f(x) = \frac{x-3}{x^2-6x+13}$
3. $f(x) = \frac{x^3-3x^2-10x}{x^2-9}$
4. $f(x) = \frac{x^2-9}{x-3}$

Answers

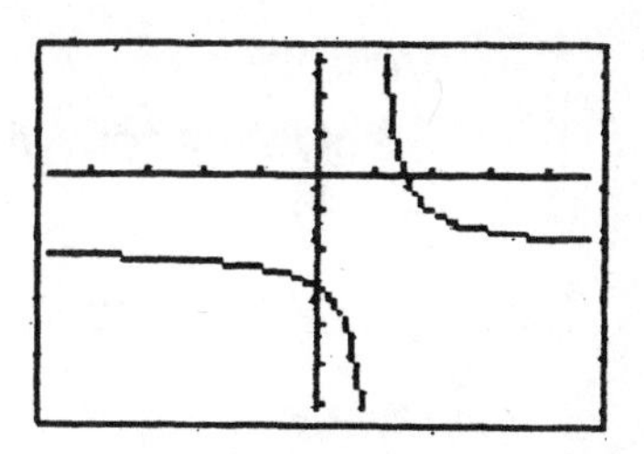

1. Window $[-4.7, 4.7]$ by $[-6.3, 3.1]$
   zeros $x = \frac{3}{2}$
   vertical asymptotes $x = 1$
   holes none
   end behavior $y = -2$

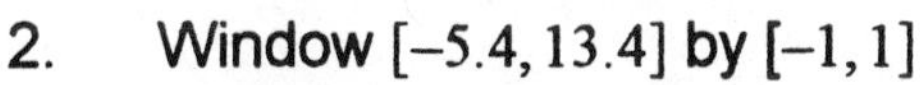

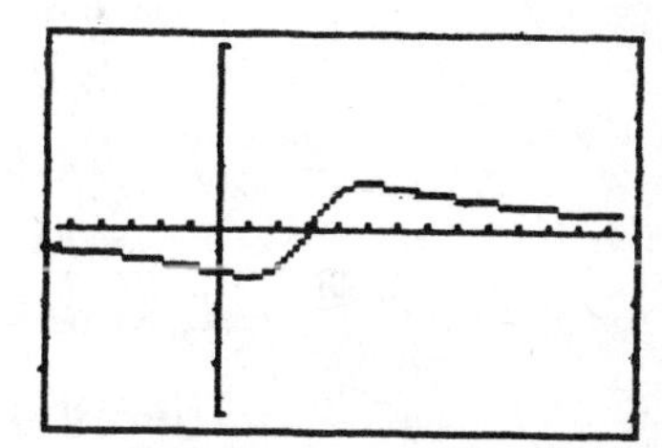

2. Window $[-5.4, 13.4]$ by $[-1, 1]$
   zeros $x = 3$
   vertical asymptotes none
   holes none
   end behavior $y = \frac{1}{x}$
   horizontal asymptote $y = 0$

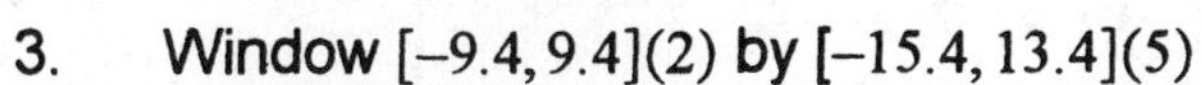

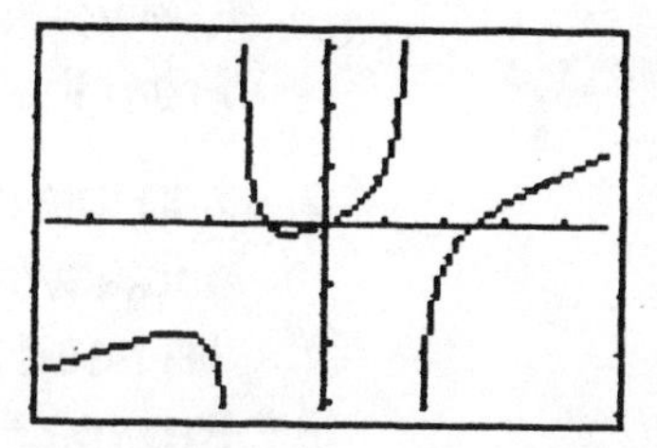

3. Window $[-9.4, 9.4](2)$ by $[-15.4, 13.4](5)$
   zeros $x = -2, 0, 5$
   vertical asymptotes $x = \pm 3$
   holes none
   end behavior $y = x$

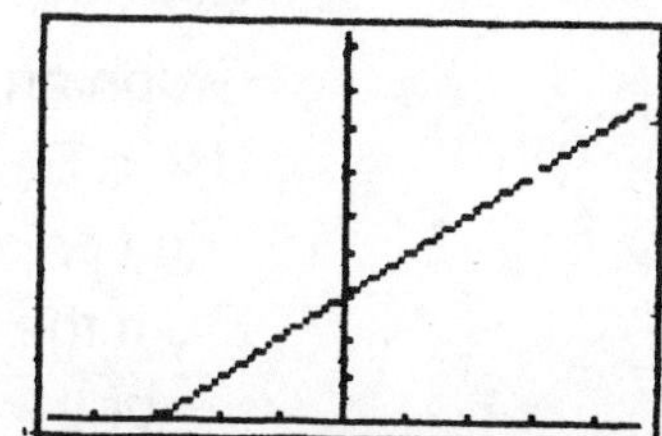

4. Window $[-4.7, 4.7]$ by $[0, 9.3]$
   zeros $x = -3$
   vertical asymptotes none
   holes none
   end behavior $y = x$

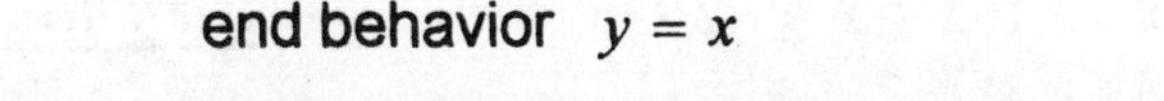

Graphs of Polynomial and Rational Inequalities

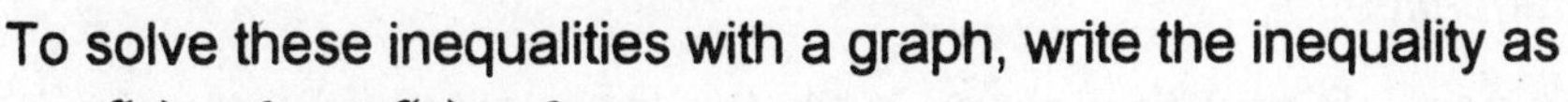

To solve these inequalities with a graph, write the inequality as
$f(x) \geq 0$, or $f(x) \leq 0$.
Graph $f(x) = 0$ on a Friendly Window.

Find the x intercepts and the vertical asymptotes for $f(x)$, these values for x are where $f(x)$ could change sign.

From the graph, read the intervals where $f(x)$ is above or below the x axis.

ex. $\frac{x-3}{x+3} \le 5$

Rewrite the inequality

$f(x) = \frac{x-3}{x+3} - 5 \le 0$

Graph using the window

[-6.7, 2.7] by [-50,50](10)

This is a Friendly Window

TRACE. the x intercept is $x = -4.5$

The vertical asymptote is $x = -3$

$f(x) \le 0$ (the graph is below the x axis)

for $x \le -4.5$ or $x > -3$

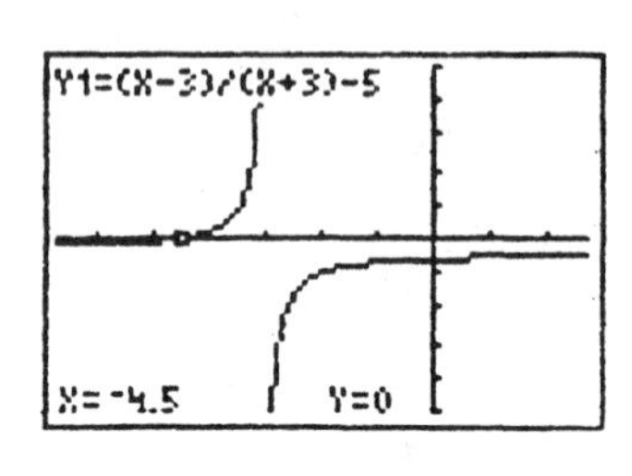

Problems

Solve the following inequalities. Find exact solutions when possible.

1. $\frac{1}{x-4} < \frac{3}{x+2}$
2. $\frac{x}{x^2-5} \le \frac{2}{x}$

Answers

1. Rewrite $f(x) = \frac{1}{x-4} - \frac{3}{x+2} < 0$

   Graph $f(x)$ on [-9.4, 9.4] by [-3.1, 3.1]

   This function has 2 vertical asymptotes and 1 intercept.

   Rewrite $f(x) = \frac{14-2x}{(x-4)(x+2)}$

   The x intercept is $x = 7$

   The vertical asymptotes are $x = -2, x = 4$

   From the graph $f(x) < 0$, for $-2 < x < 4$

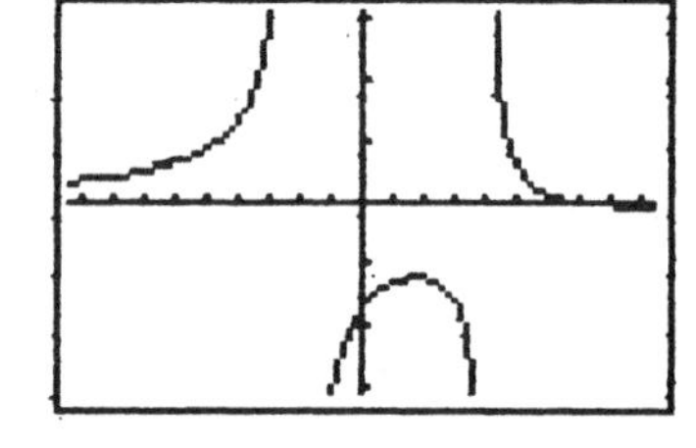

2. In this problem the values of the intercepts and asymptotes are irrational numbers. You will need to use a solver or the quadratic formula.

   Rewrite $f(x) = \frac{x}{x^2-5} - \frac{2}{x} \le 0$

   Graph on [-4.7, 4.7] by [-6.2, 6.2]

   Use a Solver for the intercepts, $x \approx \pm 3.162$

   The vertical asymptotes are at $x = 0, \pm\sqrt{5} \approx \pm 2.236$

   From the graph $f(x) \le 0$ for $-3.162 \le x \le -2.236$ or $0 < x < 2.236$ or $3.162 < x$

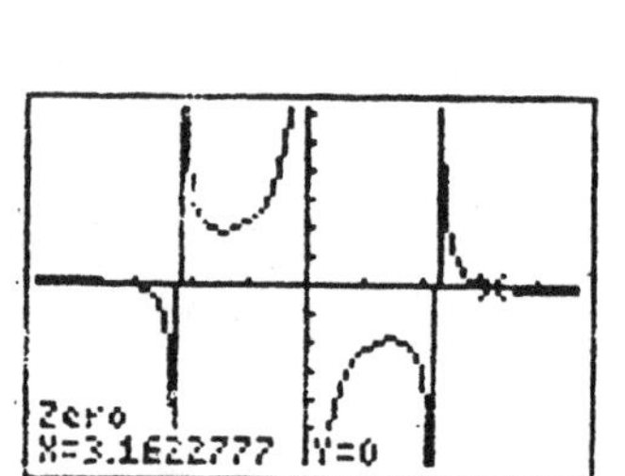

Absolute Value Inequalities

To solve these inequalities with a graph, write the inequality as $f(x) \ge 0$, or $f(x) \le 0$.

Graph $f(x) = 0$ on a Friendly Window.
Find the x intercepts for $f(x)$, these are the
values for x where $f(x)$ could change sign.
From the graph, read the intervals where $f(x)$ is above or
below the x axis.

Another method is to graph the 2 functions on the same screen.
Then determine when one graph is above (or below) the other.

Problems

Solve the following inequalities. Find exact solutions when possible.

1. $|x^2 + 2x - 9| \leq 6$
2. $|2x - 1| > 2x + 1$

Answers

1. Window [-5.7, 3.7] by [-3.1,3.1]
   Graph $f(x) = |x^2 + 2x - 9| - 6$
   Intercepts $x = -5, \ -3, 1, 3$
   $f(x) \leq 0$ for $-5 \leq x \leq -3$,
   or $1 \leq x \leq 3$

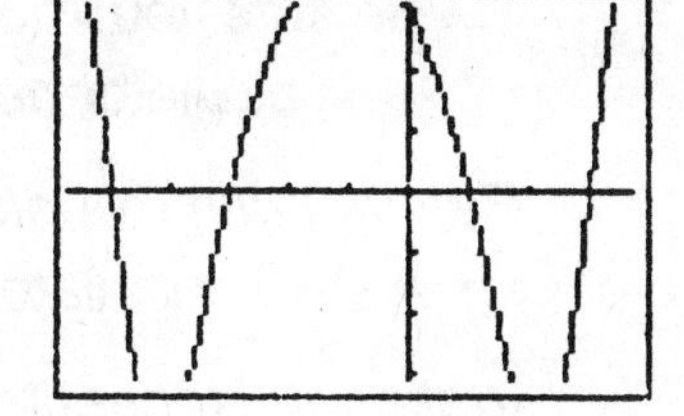

2. Decimal Window
   This time graph the 2 functions
   $y1 = |2x - 1|, \ \ y2 = 2x + 1$
   When is the absolute value graph
   above the line graph?
   The 2 graphs intersect at the point (0, 1)
   The inequality is true for $x < 0$

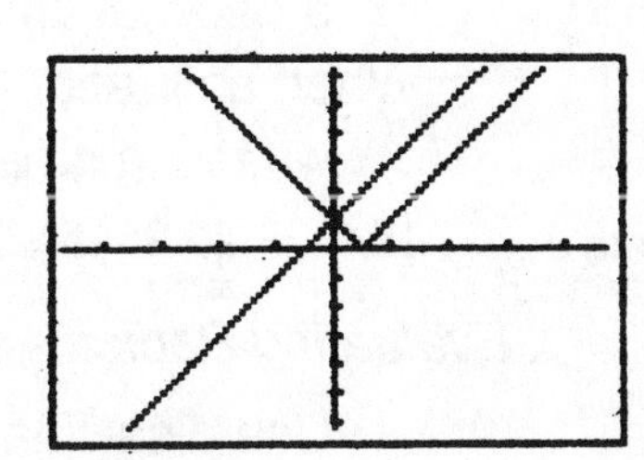

Exponential and Logarithmic Functions

Radicals and Rational Exponents

Radicals can be written as fractional exponents.

On a calculator, the nth root of c is written $\sqrt[n]{c}$ or as c^$(\frac{1}{n})$.

The value of an even root is always positive.

The value of an odd root will be either positive or negative.

Be very careful when using fractional exponents on a calculator.

When drawing a graph, the root key $\sqrt[n]{\ }$ usually will draw the correct graph.

Remember to enclose the entire argument for the radical or the entire fractional exponent in parenthesis.

ex. $(-8)^{\frac{2}{3}} = 4$, Some calculators will not display this answer. Try $\sqrt[3]{(-8)^2}$ or (-8²)^(1÷3)

Be sure you know how to enter fractional exponents on your calculator before you do the problems in this chapter.

Rationalizing Denominators

A CAS calculator, will do simple rationalization problems.

Irrational Exponents

Irrational numbers can be expressed in exact form, $\sqrt{2}$ or as a decimal approximation 1.4142135...

When an irrational number is used as an exponent, it should be written in its exact form for better accuracy.

$10^{\sqrt{2}} \approx 25.95455...$ is more accurate than 10^1.414 $\approx$ 25.9537...

Evaluating Radicals and Exponents

Use a calculator to approximate to 3 decimal places.

Problems		*Approximate Answers*
1.	$\sqrt[7]{21}$	1.545...
2.	$\sqrt[5]{e}$	1.221...
3.	$\sqrt[12]{10}$	1.212...
4.	$2^{.25}$	1.889...
5.	2^{e}	6.581...
6.	$10^{.306}$	2.023...
7.	e^{-3}	0.0498...
8.	$e^{.085}$	1.0087...
9.	$e^{\sqrt{2}}$	4.113...
10.	e^{π}	23.141...

Solving Radical Equations:

To solve radical equations with a graph, write the equation as $f(x) = 0$, and graph $f(x)$ on a Friendly Window.
Find the x intercepts for $f(x)$, these are the solutions.
The graph will show only the true solutions, not any extraneous solutions.
Another method is to graph the 2 functions on the same screen, then find the intersections. The x coordinates of the intersections are the solutions of the equation.

ex. $\sqrt{2x+7} = 2+x$

Rewrite the problem as

$f(x) = \sqrt{2x+7} - (2+x) = 0$

Graph $f(x)$ on the Decimal Window
This is a Friendly Window. TRACE
The solution is $x = 1$

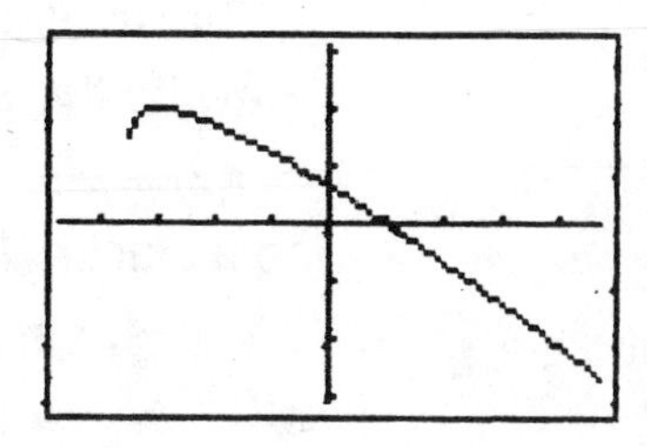

ck. Store 1 in x, evaluate $f(1)$

Problems:

Solve the following radical equations. Find exact solutions when possible.
Check all of your answers.

1. $x^{\frac{2}{3}} + 2x^{\frac{1}{3}} - 8 = 0$
2. $2x + 9\sqrt{x} = 5$
3. $x = \sqrt{11x - 30}$
4. $\sqrt{x+1} - 3x = 1$
5. $\sqrt[3]{x^3 + x^2 - 4x + 5} = x + 1$
6. $\sqrt{3x+1} - 1 = \sqrt{x+4}$

Answers

1. This problem is best solved with Algebra because the first term does not compute correctly on some calculators.
There are 2 solutions.
Graph $f(x)$ on [-75, 15](10) by [-10,10](2).
This is **not** a Friendly Window.
Use a Solver to find the intercepts,
$x = 2, x = -64$. Check these answers.

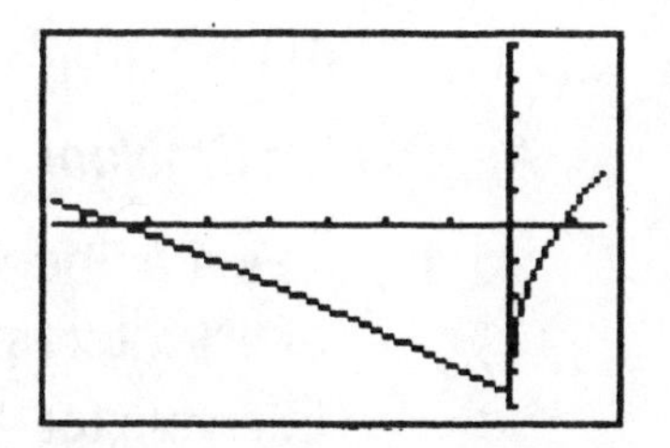

2. Rewrite the equation
$f(x) = 2x + 9\sqrt{x} - 5$
Window [-2.35, 2.35] by [-3.1, 3.1]
This is a Friendly Window. TRACE.
$x = .25 = \frac{1}{4}$

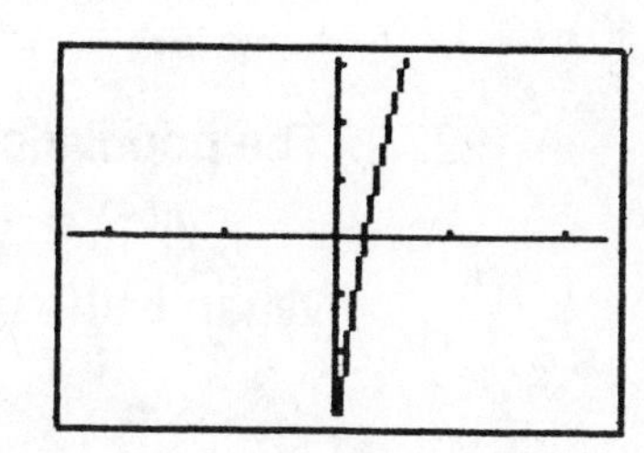

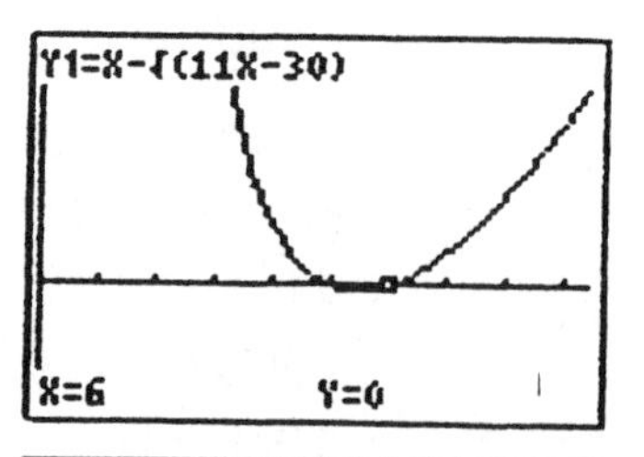

3. Rewrite the equation
   $f(x) = x - \sqrt{11x - 30}$
   Window [0, 9.4] by [-.5, 1]
   This is a Friendly Window. TRACE.
   $x = 5, x = 6$

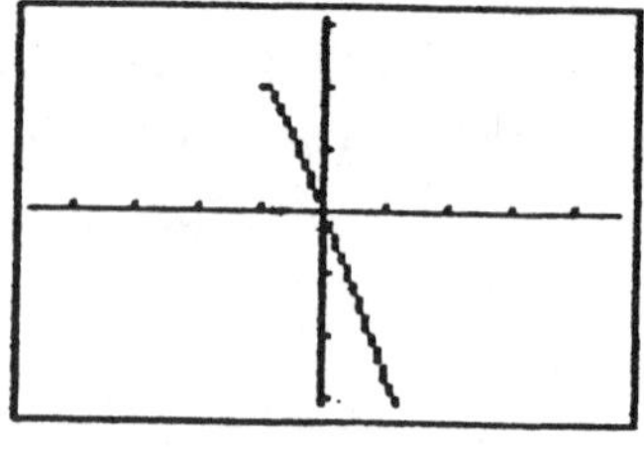

4. Rewrite the equation
   $f(x) = \sqrt{x+1} - 3x - 1$
   Use a Decimal Window, TRACE.
   The solution is $x = 0$

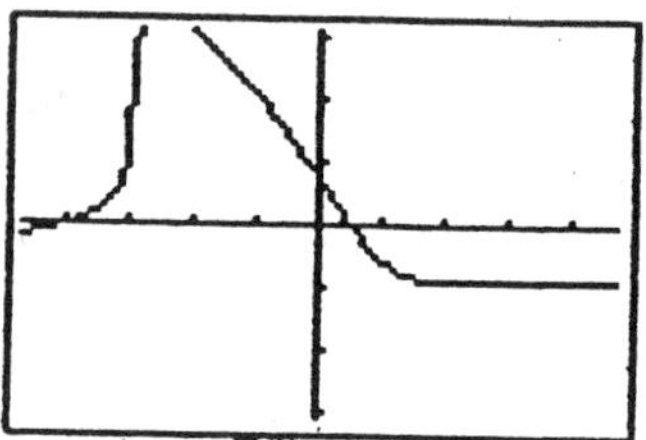

5. Rewrite the equation
   $f(x) = \sqrt[3]{x^3 + x^2 - 4x + 5} - (x + 1)$
   Use a Decimal Window, TRACE.
   $x = -4, x = .5 = \frac{1}{2}$

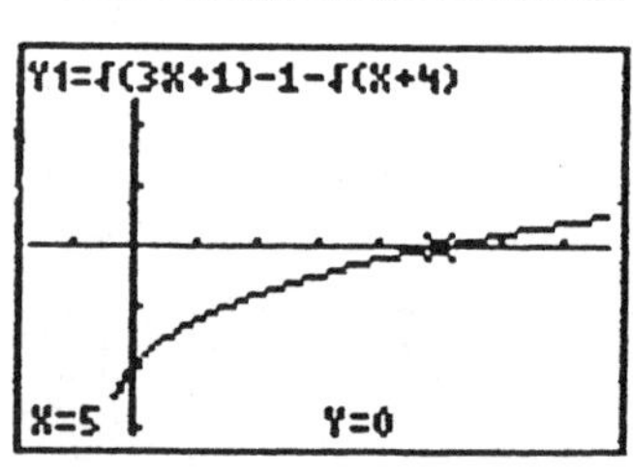

6. Rewrite the equation
   $f(x) = \sqrt{3x+1} - 1 - \sqrt{x+4}$
   Window [–1.7, 7.7] by [–3.1, 3.1]. TRACE.
   $x = 5$

Exponential Functions

The domain of $y = a^x$ is all real x, the range is $y > 0$.
The graph of a^x for $a > 0$ is always positive and always increasing
The graph of a^x for $a > 0$ is always negative and always decreasing.
When using the e^x or the 10^x keys, do not use ^. Press e^x or 10^x, then enter the exponent (in parenthesis if it is a function).
There are many applications that are modeled with exponential functions.
A complete graph of $f(x) = c^x$ is on $[a, b]$ by $[f(a), f(b)]$

Application Problem

1. Find a Friendly Window to model the current value of $200 at 5% interest so that TRACE shows the values at $\frac{1}{2}$ year intervals for 40 years.

2. The population of fish in a lake at time x in months is
   $p(x) = \frac{10{,}000}{1+19e^{(-x/5)}}$
   When is the population of fish = 2,000?

3. Enter the following data in the STAT menu.
   x is the number of minutes of Advertising on TV
   y is the profit in sales for the next week in ten thousand dollars.

x	1	2	3	4	5	6
y	12.0	10.8	15.0	22.0	22.6	28.9

   Find an Exponential Regression that approximates the data.
   Draw the Scatter Plot and the Regression Line.

Answer

1. The equation is $f(x) = 200(1.05)^x$
   Compute $f(0) = 200, f(40) \approx 1410$
   A Friendly Window containing these values is [0, 47](5) by [100, 2000](100)
   As Xmax - Xmin = .5(96) Each pixel is .5
   TRACE. The x values are at $\frac{1}{2}$ year intervals.
2. You are looking for the value of x that satisfies
   $p(x) = \frac{10,000}{1+19e^{(-x/5)}} = 2000$
   Rewrite the equation $f(x) = 1 + 19e^{(-x/5)} - 5$
   Graph on the Friendly Window
   [0, 9.4](2) by [-3.1, 3.1]
   TRACE does not give an exact answer,
   so use a Solver
   $x \approx 7.791$ or about 8 months.

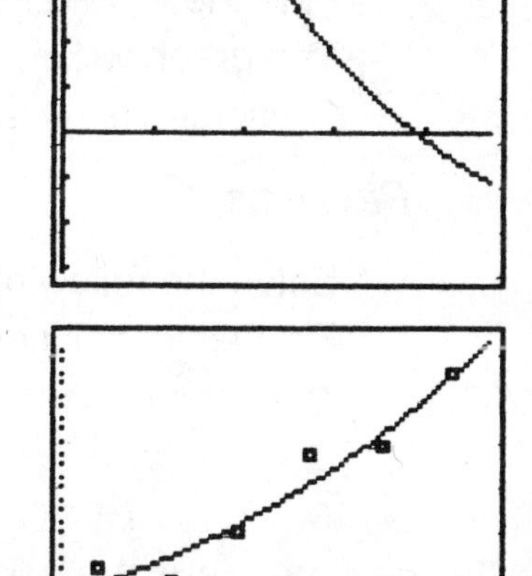

3. The regression equation is
   $y = 8.248(1.251)^x$

Logarithms

The logarithm functions are the inverse functions of the exponential functions.
The domain of $y = \log(x)$ is $x > 0$, the range is all real y.
$y = 10^x$ has the inverse function $y = \log(x) = \log_{10}(x)$.
$y = e^x$ has the inverse function $y = \ln(x) = \log_e(x)$.
$y = 2^x$ has the inverse function $y = \log_2(x)$.
The only logarithms that have keys on a calculator are $\log(x)$ and $\ln(x)$.
To find logarithms to another base b, use $\log_b N = \frac{\ln N}{\ln b}$
To check that these functions are inverses, Press

$e^{\ln(2)} = \underline{2}$, $10^{\log(5)} = \underline{5}$

A logarithm is an exponent. The following equations are equivalent. $\log_b N = L \Leftrightarrow b^L = N$

Evaluating Logarithms

Use a calculator to approximate to 3 decimal places.

Problems		*Approximate Answers*
1.	$\ln(\frac{5}{2})$	.916
2.	$\ln(e^2)$	2
3.	$\ln(5^2)$	4.828
4.	$\log(0.02315)$	– 1.635
5.	$\log_2(39.5) = \frac{\ln(39.5)}{\ln(2)}$	5.304
6.	$\log_{\frac{1}{4}}(\frac{1}{3}) = \frac{\ln(\frac{1}{3})}{\ln(\frac{1}{4})}$	.7925

Solving Exponential and Logarithmetic Equations.

To solve radical equations with a graph,
Write the equation as $f(x) = 0$, and
Graph $f(x)$ on a Friendly Window.
Find the x intercepts for $f(x)$, these are the solutions.
The graph will show only the true solutions, not any extraneous solutions.

Problems

Solve the following exponential and logarithmic equations.
Find exact solutions when possible.

1. $5^x = 2$
2. $\ln(x) = 2$
3. $y = e^{2x} - 3e^x + 2$
4. $2^{(4x-1)} = 3^{(1-x)}$
5. $e^x - e^{-x} = 4$
6. $\ln\sqrt{x+2} = 1$
7. $\ln(x) + \ln(x+2) = 3$
8. Find the local maximum and minimum points on the graph of $y = xe^{(-x^2)}$.

Answers

Most of these answers are irrational. You will need to use a Solver
If you TRACE, then use a numeric solver, the value of the last point on Trace is entered automatically as the guess.

1. Rewrite the equation

   $f(x) = 5^x - 2$

   Use a Decimal Window. TRACE

The answer is not exact.
Use a Graphic or Numeric Solver
$x \approx .431...$

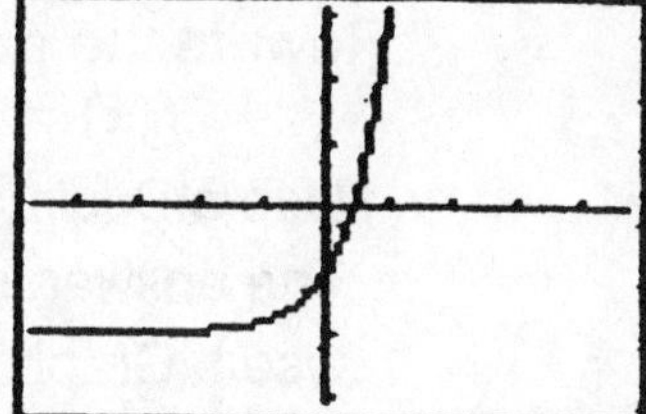

2. Rewrite the equation
$f(x) = \ln(x) - 2$
Use Window [0, 9.4] by [-3.1, 3.1] TRACE
The answer is not exact.
Use a Graphic or Numeric Solver
The solution is $x \approx 7.389...$

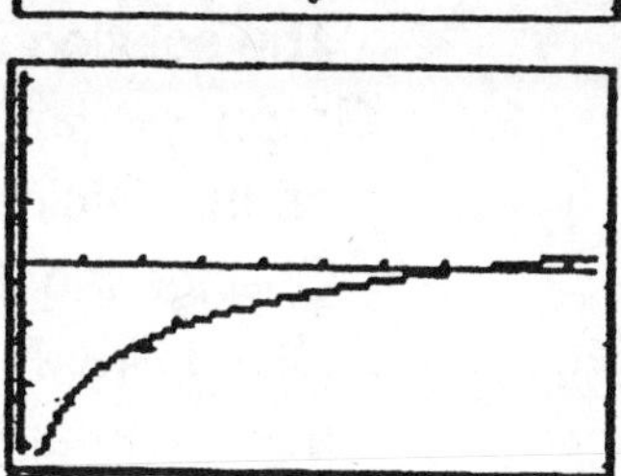

3. Rewrite the equation
$f(x) = e^{2x} - 3e^{x} + 2$
Window [-2.35,2.35] by [-3.1,3.1]
TRACE. One solution is $x = 0$
The other answer is not exact.
Use a Graphic or Numeric Solver
The solution is $x \approx .693...$

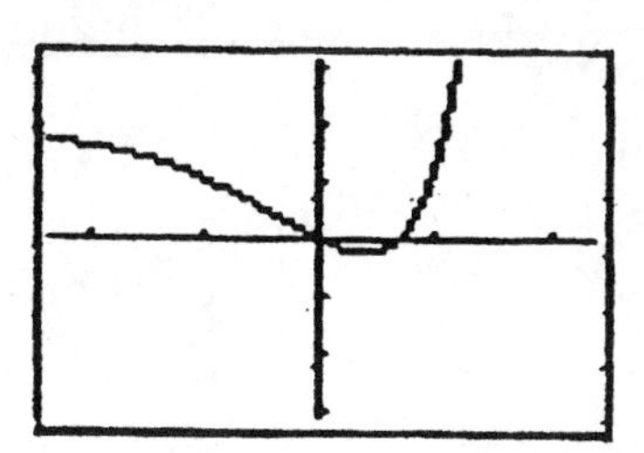

4. Rewrite the equation
$f(x) = 2^{(4x-1)} - 3^{(1-x)}$
Use a Decimal Window.
Change X Range to [0, 9.4]. TRACE
The answer is not exact.
Use a Graphic or Numeric Solver
The solution is $x \approx .463...$

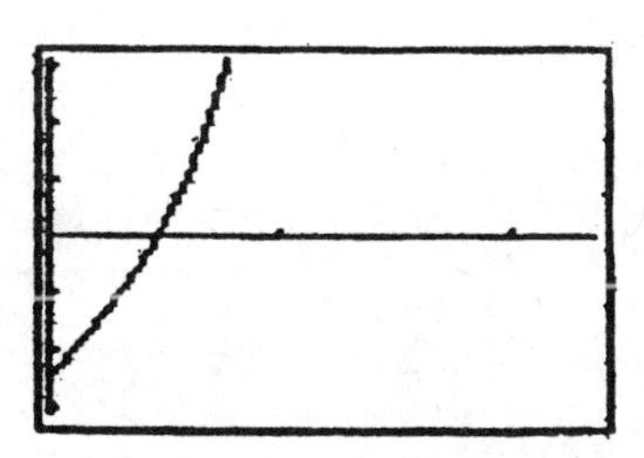

5. Rewrite the equation
$f(x) = e^{x} - e^{-x} - 4$
Use a Decimal Window. TRACE
The answer is not exact.
Use a Graphic or Numeric Solver
The solution is $x \approx 1.444...$

6. Rewrite the equation
$f(x) = \ln\sqrt{x+2} - 1$
Use a Decimal Window.
TRACE. The answer is not exact.
Use a Graphic or Numeric Solver
The solution is $x \approx 5.389...$

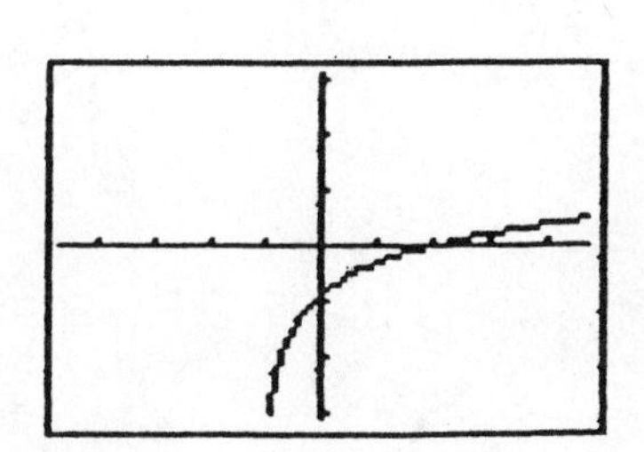

7. Rewrite the equation

$$f(x) = \ln(x) + \ln(x+2) - 3$$

Use a Decimal Window. TRACE

The answer is not exact.

Use a Graphic or Numeric Solver

The solution is $x \approx 3.592...$

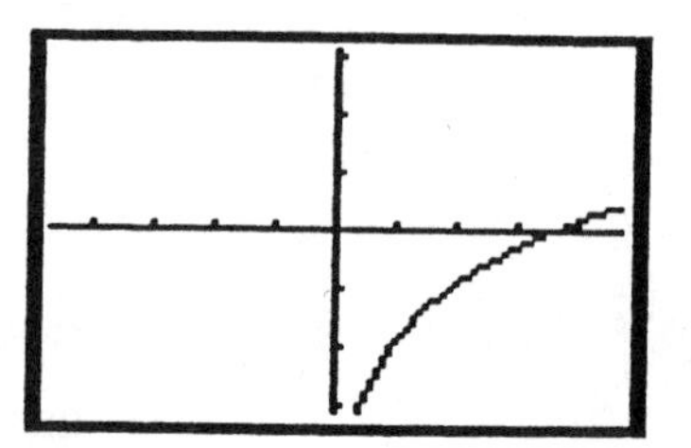

8. Graph $y = xe^{(-x^2)}$

on the Decimal Window.

Change the YRange to [-1, 1]

Use a Graph Solve to find the maximum and minimum values

The max value is $y \approx .4289...$ at $x \approx .707...$

The min value is $y \approx -.4289...$ at $x \approx -.707...$

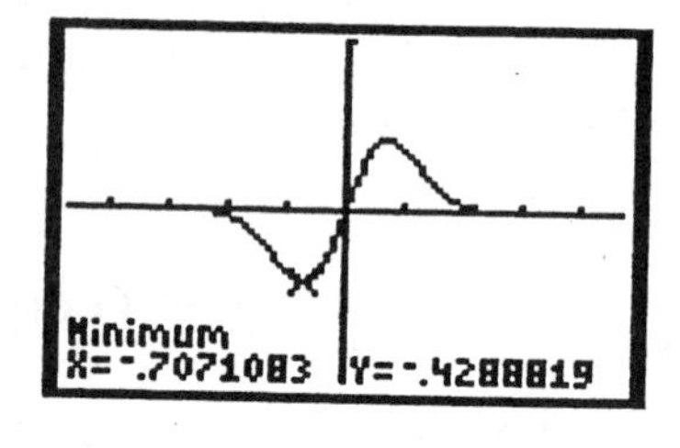

Trigonometric Functions

Angle Measurement:

Angles are measured in either degrees or radians.
One full revolution of an angle through a circle is 360° or 2π radians.
To change between degrees and radians, π radians = 180°
It is easy to use a proportion to solve these problems.
The symbol π is entered from the keyboard. The calculator will change it to the appropriate decimal when it is used in calculations.
The textbook uses t to represent the angle (input) in trigonometric functions, however to graph a function, some calculators require the input angle to be written using the variable x.

$$\frac{\text{number of radians}}{\text{number of degrees}} = \frac{\pi}{180°}$$

ex. Change 45° to radians

$$\frac{x}{45} = \frac{\pi}{180}$$
$$x = \frac{\pi(45)}{180}$$
$$x = \frac{\pi}{4}$$

ex. Change $\frac{\pi}{2}$ to degrees

$$\frac{\frac{\pi}{2}}{x} = \frac{\pi}{180}$$
$$x = \frac{180\pi}{2\pi}$$
$$x = 90°$$

To do **all** calculations in either MODE, set either Degree or Radian in the MODE menu or SET UP screen.
On all calculators, you may use the degree symbol to enter an angle in degrees while in radian MODE, and use the radian symbol to enter an angle in radians while in degree MODE
ex. The calculator is set in Radian MODE.
sin(30°) = .5
For the problems in Chapter 6, set Radian in the MODE menu.

Entering Trigonometric Functions:

The functions $\sin(x)$, $\cos(x)$, and $\tan(x)$ are on the calculator keyboard.
Some calculators force a parenthesis, if yours doesn't, use parentheses to enclose the function input.
Always end with a parenthesis even if the calculator doesn't require it.

To raise a trigonometric function to a power, enter $(\sin(x))^2$.
The exponent -1 is **not** used to write $\frac{1}{\sin(x)}$.
The function $\tan(x)$ can be entered $\frac{\sin(x)}{\cos(x)}$, the end parenthesis
is needed here.
The other functions are entered using the definitions
$\sec(x) = \frac{1}{\cos(x)}$, $\csc(x) = \frac{1}{\sin(x)}$ and $\cot(x) = \frac{1}{\tan(x)} = \frac{\cos(x)}{\sin(x)}$.

Dynamic Graph of the sine and cosine:

Change the calculator to Parametric Mode.
Change SEQUENTIAL to SIMULTANEOUS
Set Window [-3, 2π] by [-3, 3]
Enter $x1 = \cos(T)$, $y1 = \sin(T)$, $x2 = T$, $y2 = \sin(T)$ Graph
Both equations are graphed at the same time. The y values
are identical at all times.
On some calculator screens, the circle is not round. (The true shape)
Regraph on a SQUARE window.

Setting the Window for Trig Functions:

As the period 2π shared by all of the trigonometric functions is an
irrational number, it is very difficult to find a Friendly Window.
If the angle is in degrees, it is possible to have each pixel 1°.
Set Xmax - Xmin = k (number of pixels on the length of the
screen) where Xmax - Xmin > 360. (The constant k can have
decimal values).
The best way to view Trigonometric functions is to use the Trig
Window built into the calculator.

Graphs of the Trigonometric Functions:

(the vertical lines are technology errors)

$y = \sin(x)$

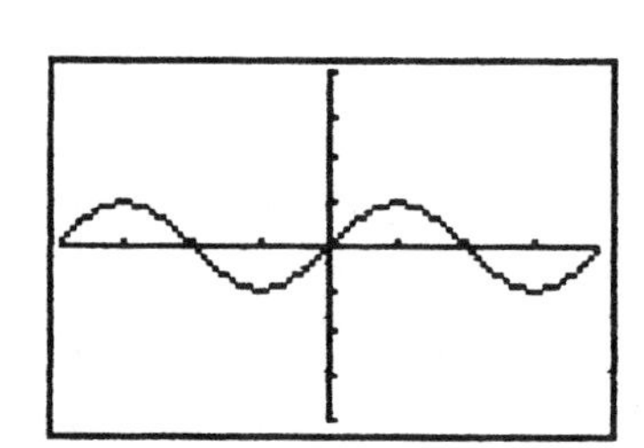

$y = \cos(x)$

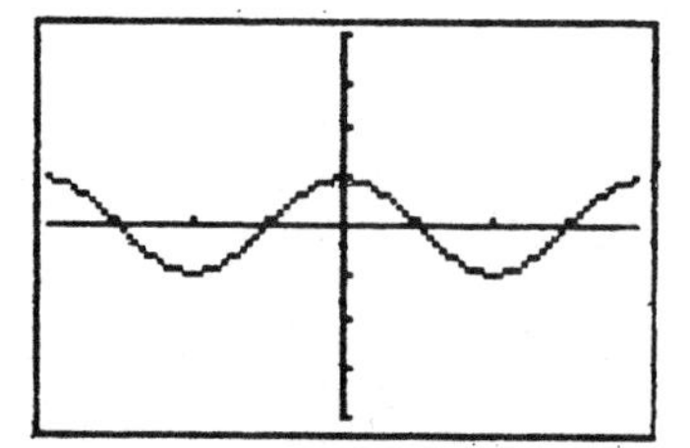

$y = \tan(x)$

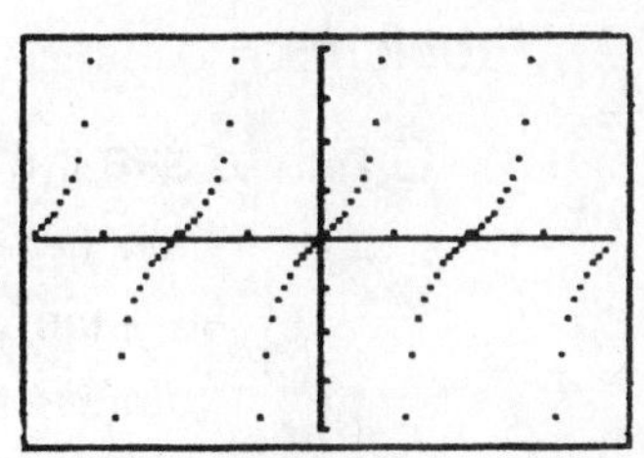

$y = \sec(x) = \frac{1}{\cos(x)}$

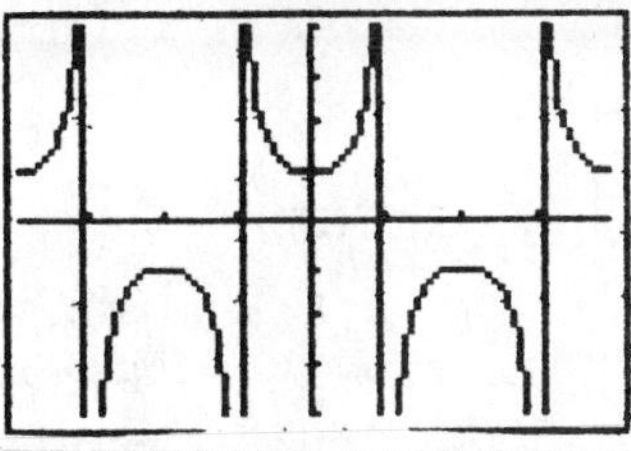

$y = \cot(x) = \frac{\cos(x)}{\sin(x)}$

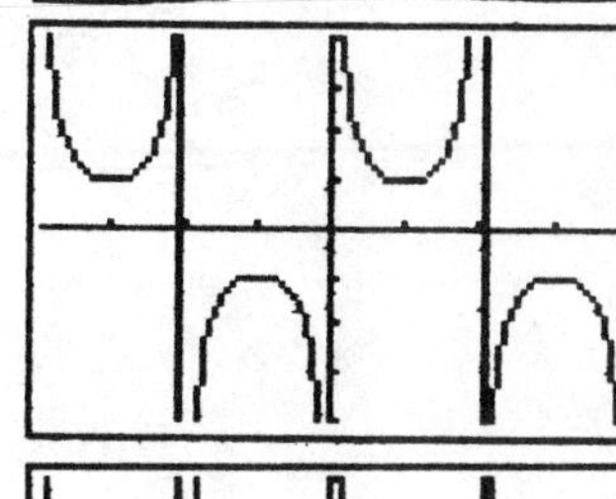

$y = \csc(x) = \frac{1}{\sin(x)}$

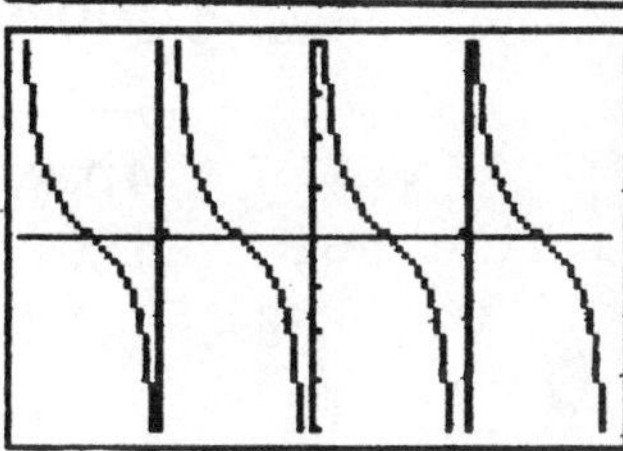

Is this an identity?

If 2 functions are graphed on the same screen, and they are indistinguishable, they may be identical. To be certain, they must be investigated Algebraically.

If 2 functions are graphed on the same screen and they are not the same, it is obvious that they are not identical.

ex. Is $\sin(2x) = 2\sin(x)$?

Set your calculator to Radian Mode.

Enter $y1 = \sin(2x), y2 = 2\sin(x)$

Graph on the Trig Window

The functions are not the same.

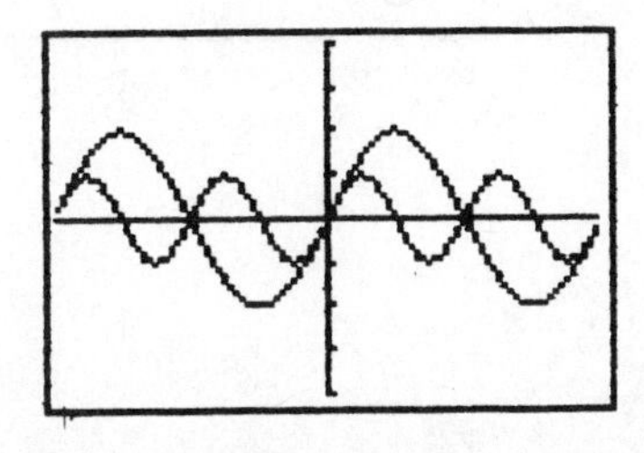

both graphs

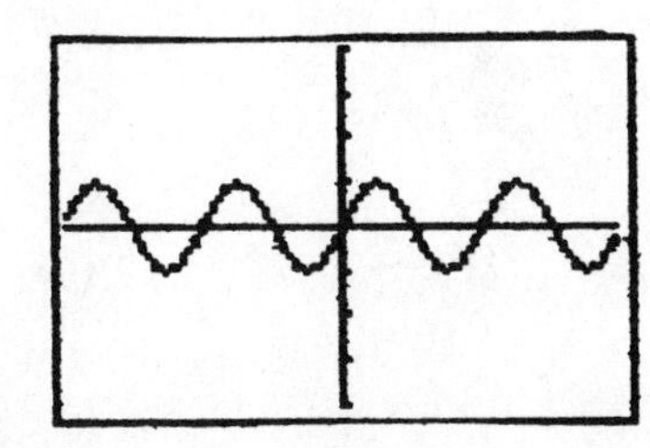

$\sin(2x)$

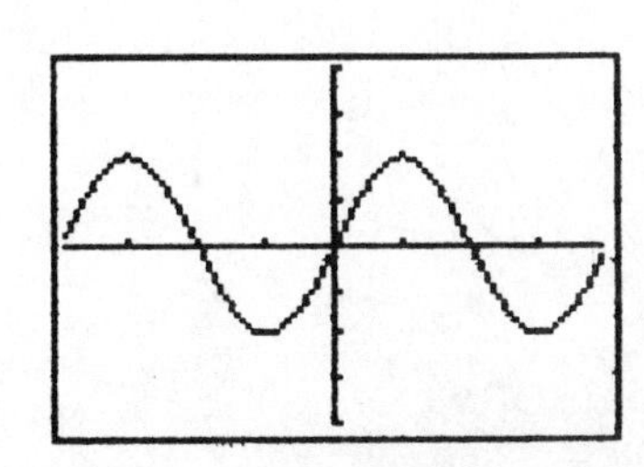

$2\sin(x)$

Problems:

Graph to see if the functions are the same.

1. $\sin^2(x) + \cos^2(x) = 1$
2. $1 + \tan^2(x) = \sec^2(x)$
3. $\sin(x + \pi) = \sin(x) + \sin(\pi)$
4. $\sin(x + 2\pi) = \sin(x)$

Answers:

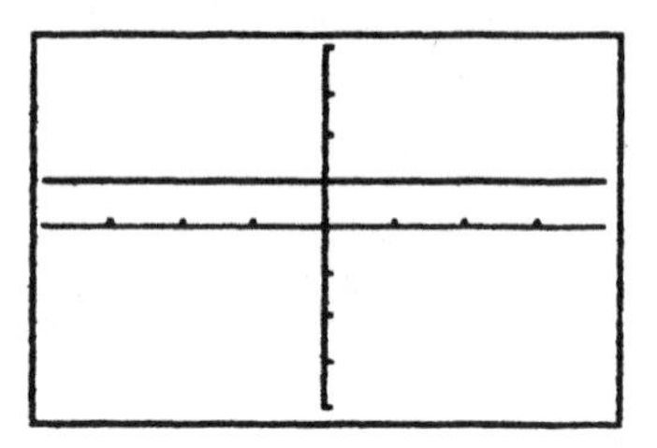

1. They appear the same.
   Prove using the Pythagorean Theorem.
   They are the same.

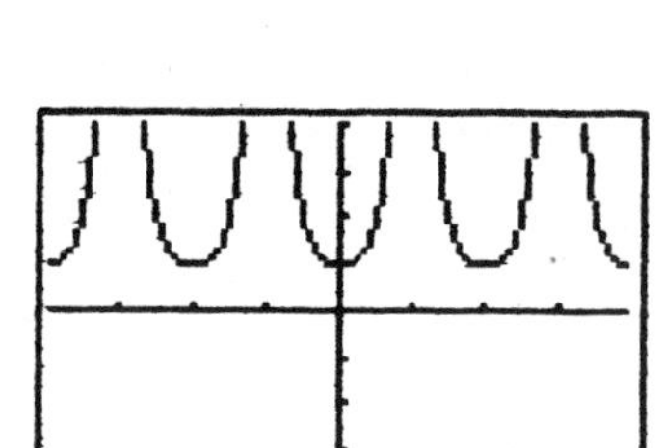

2. They appear the same.
   Prove using problem 1.and
   the definitions of sec and tan.
   They are the same.

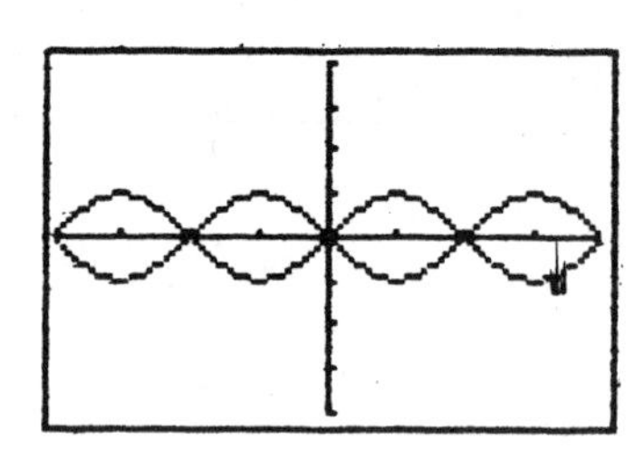

3. They are not the same.

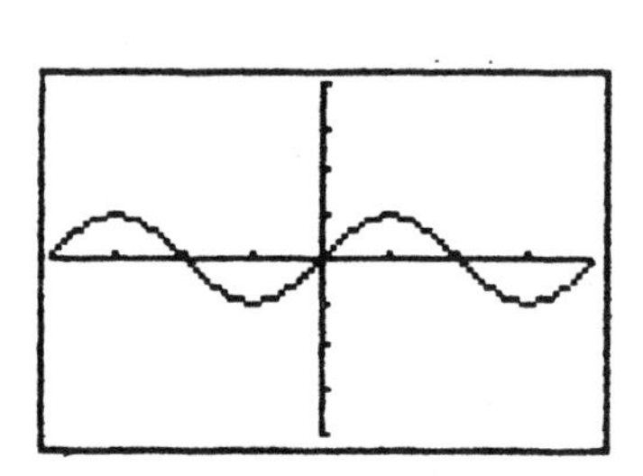

4. They appear the same.
   They are the same by the
   definition of periodic functions.

Chapter 2 – Calculator Notes and Problems

Evaluating Trigonometric Functions:

Use a calculator to evaluate to 3 decimal places. If no unit is given, the angle is in radians.

	Problems	Approximate Answers
1.	$\sin 46°$	0.719
2.	$\tan 108°$	-3.078
3.	$\cos 325°$	0.819
4.	$\cot 202°$	2.475
5.	$\sec 76°$	4.134
6.	$\sin 2$	0.909
7.	$\cos\left(\frac{\pi}{15}\right)$	0.978
8.	$\tan \pi$	0
9.	$\sec\left(\frac{1}{2}\right)$	1.139
10.	$\tan .75$	0.932
11.	$\cos\left(\frac{11\pi}{6}\right)$	0.866
12.	$\sin\left(\frac{-13\pi}{4}\right)$	0.707

Basic Properties of Trigonometric Graphs:

Given the equation $y = A\sin(Bx + C)$

The constant A is the *amplitude*. On a graph, Ymax – Ymin > $2A$.

The constant B determines the *period* = $\frac{2\pi}{B}$.

On a graph, Xmax – Xmin > $\frac{2\pi}{B}$

The constant C is the *phase shift*.

Subtract C from both Xmax and Xmin to translate the graph.

Problems:

1. Graph $y = A\sin(x)$ for $A = 3, 2, 1, -1, -2, -3$
   What happens to the graph as A changes?
2. Graph $y = \sin(Bx)$ for $B = 3, 2, 1, -1, -2, -3$
   What happens to the graph as B changes?
3. Graph $y = \sin\left(x + \frac{\pi}{C}\right)$ for $C = 3, 2, 1, -1, -2, -3$
   What happens to the graph as C changes?

Answers:

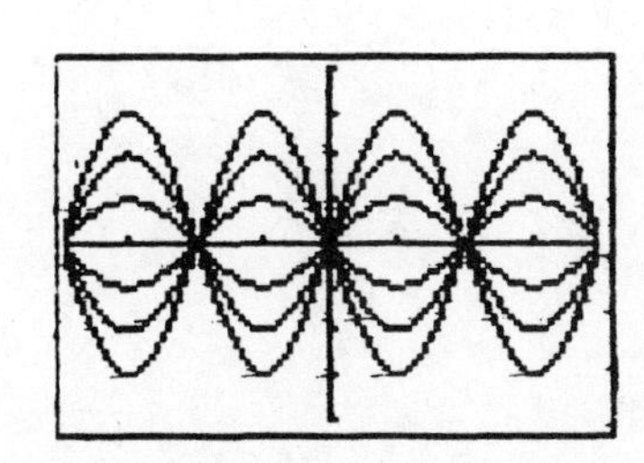

1. The amplitude changes.
   The larger the amplitude,
   the larger the maximum value of y.

2. The period changes.
   The larger the period,
   the more complete cycles fit
   into 2π radians on the X-axis.

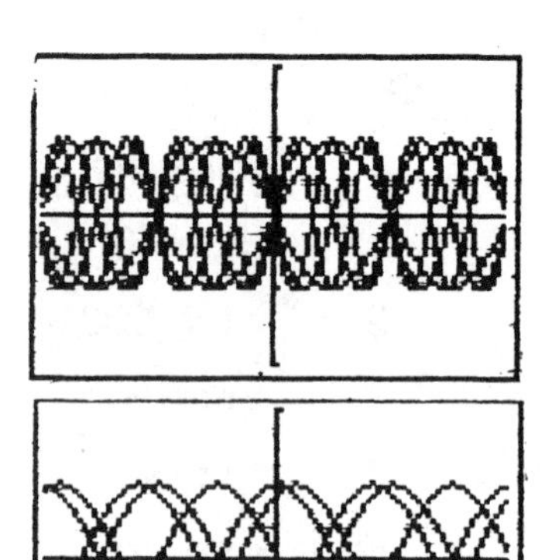

3 The phase shift changes.
   The entire curve is translated
   along the x-axis.

Finding a Complete Graph:

ex. Find a complete graph showing one complete period
   $y = 3\sin(2x - \pi)$
   Graph the function in the default trig window.
   The amplitude is 3, so there must be
   at least 6 units on the y axis.
   Rewrite the equation $y = 3\sin(2(x - \frac{\pi}{2}))$
   The period is $\frac{2\pi}{2} = \pi$.
   The equation $y = 3\sin(x)$ is shifted by $\frac{\pi}{2}$units.
   When $x = \frac{\pi}{2}, y = 0$, so add $\frac{\pi}{2}$ to the Xmin and Xmax values.
   The window is $[\frac{-\pi}{2}, \frac{3\pi}{2}]$ by $[-3.5, 3.5]$

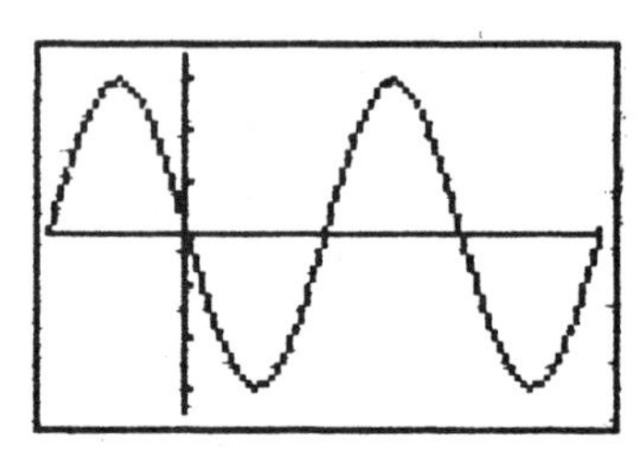

ex. $y = e^{\sin(x)}$
   $\sin(x)$ has a period of 2π
   so the default trig window
   will show a complete graph.

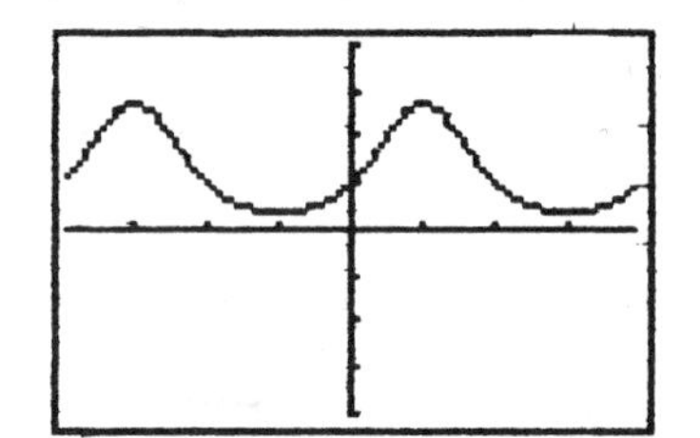

ex. $y = \ln|\sin(x) + 1|$
   $\sin(x)$has a period of 2π
   so the default trig window
   will show a complete graph.

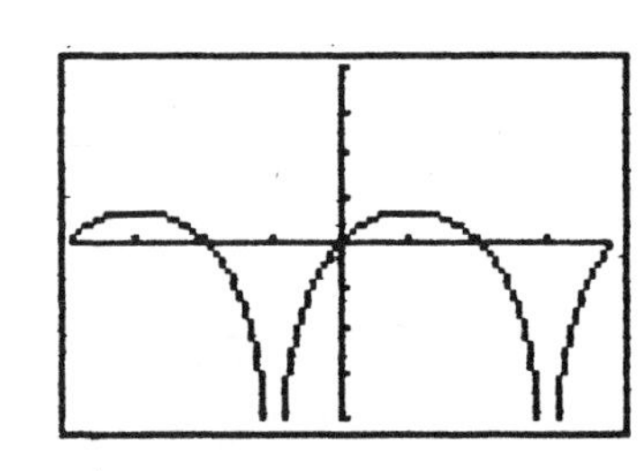

Chapter 2 – Calculator Notes and Problems

Problems:

1. Use a graph to determine the number of solutions on the interval $0 \le x \le 2\pi$. Where (what quadrants) are the solutions?
   Hint: Write the functions as $f(x) = 0$,
   a.) $\sin(x) = \frac{1}{2}$
   b.) $2\tan(x) = 4$
   c.) $\cos(3x) = \frac{1}{\sqrt{2}}$

2. Find a window for a complete graph of $y = -2\cos(3x + \frac{\pi}{2})$

Answers:

1a.) Window $[0, 2\pi]$ by [-1.5, 1.5]
2 in Q1 and Q2

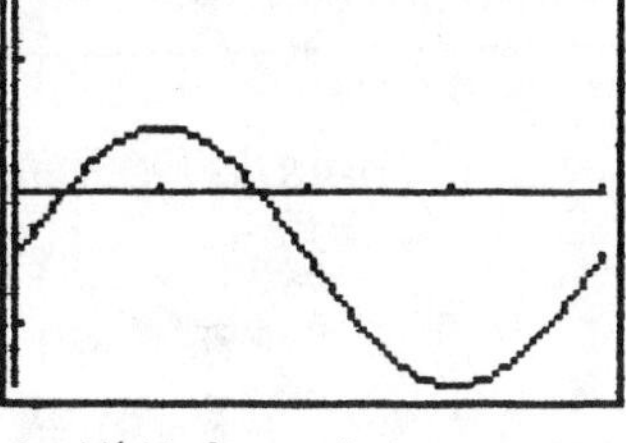
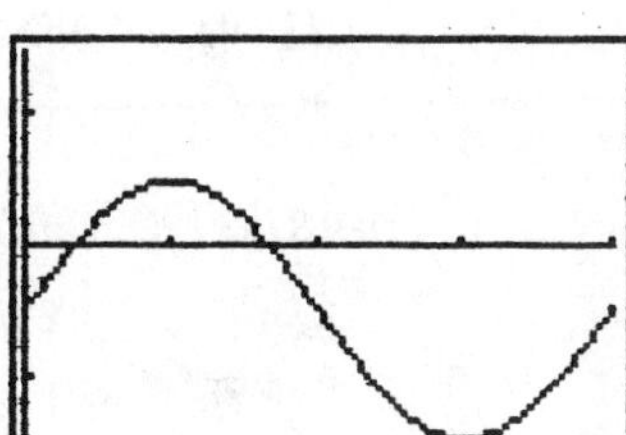

1b.) Window $[0, 2\pi]$ by [-5, 5]
2 in Q1 and Q3,
the vertical lines are technology errors.

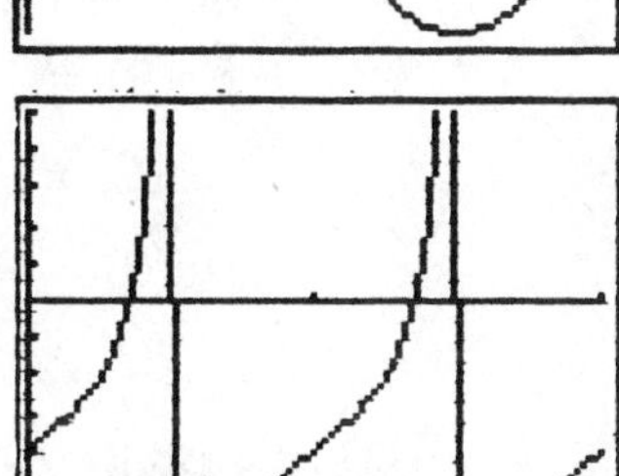

1c.) Window $[0, 2\pi]$ by [-2, 2]
6 in1 in Q1, 2 in Q2,
2 in Q3, 1 in Q4.

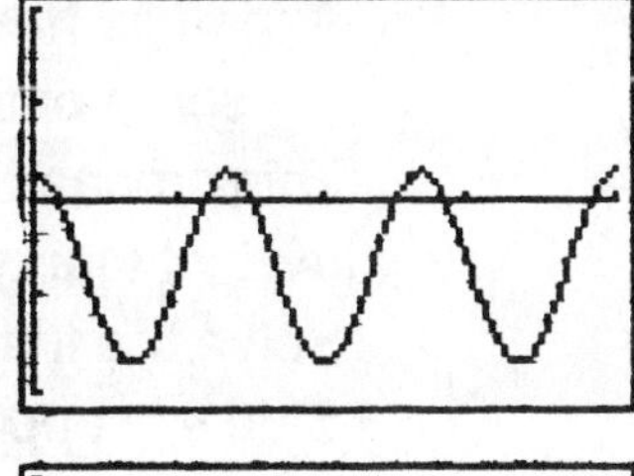

2. Amplitude = 2,
Period = $\frac{2\pi}{3}$,
Phase Shift = $\frac{\pi}{6}$,
Window $[0, \frac{2\pi}{3}]$ by [-3, 3]

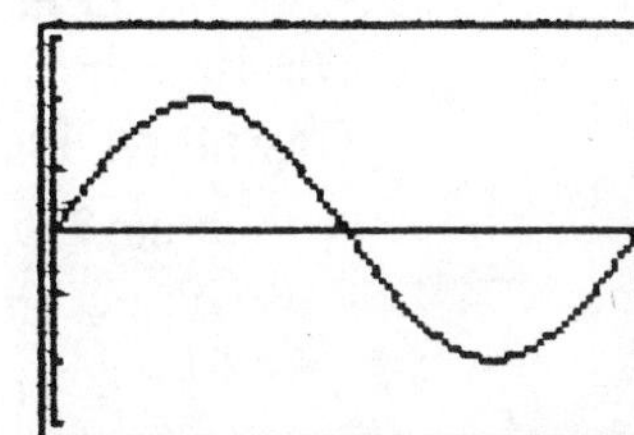

Triangle Trigonometry

Triangle Trigonometry:

Change the angle measurement to Degrees in the MODE or SET UP Menu.
For all of the problems in Chapter 7, use Degrees .
The textbook uses θ to represent the angle (input) in trigonometric functions, however to graph a function, some calculators require the input angle to be written using the variable x.
The exact values of the special angles 30°,45°,60° should be memorized.
The trigonometric functions of these angles and multiples of them are found using pencil and paper. The functions of any other angles can be found using a calculator.

Triangle Definitions of the Trigonometric Functions:

Let θ be an acute angle in a right triangle. The trigonometric functions can be defined as follows:

$$\sin(\theta) = \frac{\text{opposite side}}{\text{hypotenus}}$$
$$\cos(\theta) = \frac{\text{adjacent side}}{\text{hypotenus}}$$
$$\tan(\theta) = \frac{\text{opposite side}}{\text{adjacent side}}$$

Solving Right Triangles:

A triangle has 3 sides: a, b, c and 3 angles A, B, C. When the measures of some of these quantities are given, it is possible to find all of the others.
This process is called "solving a triangle."

ex. A right triangle has $C = 90°, A = 75°$ and $c = 17$.
Solve the triangle.
Side a is opposite the angle $A = 75°$.
The hypotenuse $c = 17$ is opposite the right angle $C = 90°$
Therefore find a, b, A.

$$\sin 75° = \frac{a}{17} \qquad \cos 75° = \frac{b}{17}$$
$$a = 17\sin 75° = 16.42 \qquad b = 17\cos 75° = 4.40$$
$$B = 90° - 75° = 15°$$

ex. Use a graph and a Solver to find θ if $\cos(\theta) = .8$
Graph $y = \cos(x) - 0.8$ using a window [0, 90]scl15 by [-1, 1]scl .5
The graph crosses the axis between $x = 30°$ and $x = 45°$
Using these numbers as bounds, use a Solver to find $x = 36.87°$

ex. Batman is on the edge of a 200 foot chasm and wants to jump to the other side.
A tree on the edge of the chasm is directly across from him.
He walks 20 feet to his right and notes that the angle to the tree is now 54°.
His jet belt allows him to jump a maximum of 24 ft.

How wide is the chasm and is it safe for Batman to jump?

$\tan(54°) = \frac{x}{20}$

$x = 20\tan(54°) = 27.528ft.$

No! It is not safe for him to jump.

ex. Two boats lie on a straight line with the base of a lighthouse. From the top of the lighthouse, (21 meters above water level) it is observed that the angle of depression of the nearest boat is 53° and the angle of depression of the farthest boat is 27°
How far apart are the prows of the boats?

Find the angles adjacent to the side and top of the lighthouse.
The angle to the nearest boat is $90° - 53° = 37°$.
The angle to the farthest boat is $90° - 27° = 63°$.
These triangles are right triangles.
The distance between the boats

$x = b_2 - b_1 = 21(\tan(63°) - \tan(37°)) = 25.39meters.$

Using the Inverse Function Keys:

Instead of solving the equation $\cos(\theta) = 0.8$ from a graph, the value of θ can be found by using the inverse cosine key. The question asked is " What angle has a cosine of 0.8?" Actually there are lots of angles that have $\cos(\theta) = 0.8$. This chapter is considering right triangles, so the problem is to find an angle $\theta < 90°$ that has $\cos(\theta) = 0.8$. The inverse function can be used to find this angle θ. This key is labeled $\cos^{-1}$, $a\cos$, or $\arccos$.
Solve the problem by pressing $\cos^{-1}(.8)$. The answer is 36.87°.

There are keys for inverse sin, $\sin^{-1}$,

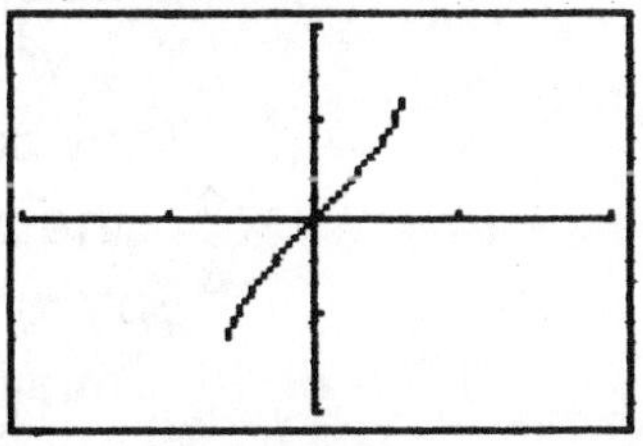

inverse cos, $\cos^{-1}$,

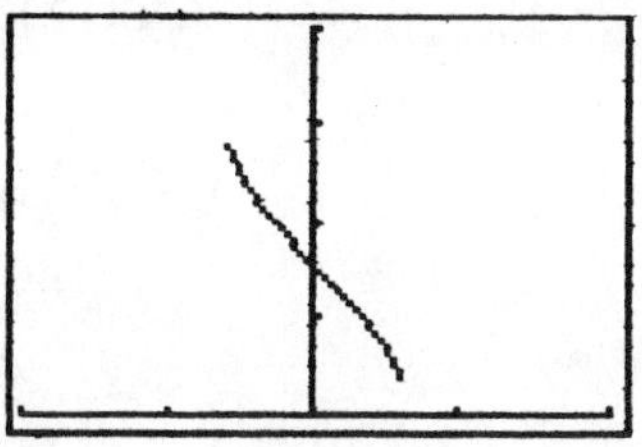

inverse tan, $\tan^{-1}$

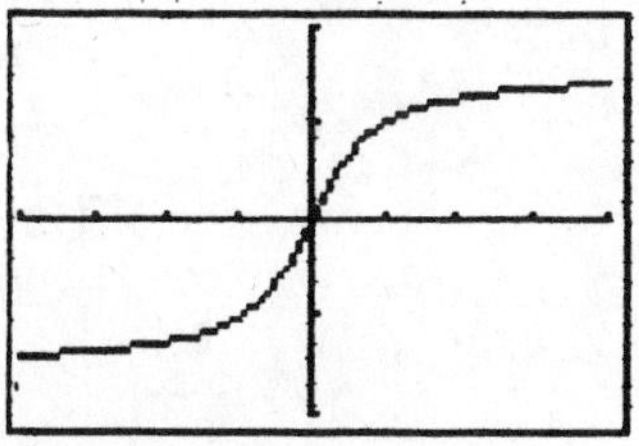

as shift keys on your calculator.

Remember, **the inverse trigonometric function is an angle.**

Trigonometric Identities and Equations

Trigonometric Identities:

Identities are equations that are valid for all values of the variables for which the equation is defined.

To prove that an equation is an identity, it is usually necessary to prove algebraically that one side can be changed into the other.

To prove that an equation is not an identity, it is sufficient to show that there is at least one value of the variable where the sides of the equation are not equal.

A graphing calculator can help in deciding if an equation is **not** an identity. If the graphs of both sides of the equation do not coincide on some complete window, the equation is not an identity.

However, if the graphs of both sides of the equation do coincide in some window, the equation may or may not be an identity. You do not know. It is impossible to see all of the values of the variable on any calculator window, and some values are too large or too small to be graphed by the calculator.

ex. Graph the following equations to determine if they could be identities.

a.) $\frac{1+\sin(x)-(\sin(x))^2}{\cos(x)} = \cos(x) + \tan(x)$

Enter the left side of the equation as Y1, enter the right side of the equation as Y2. If possible on your calculator, use different displays for each function.

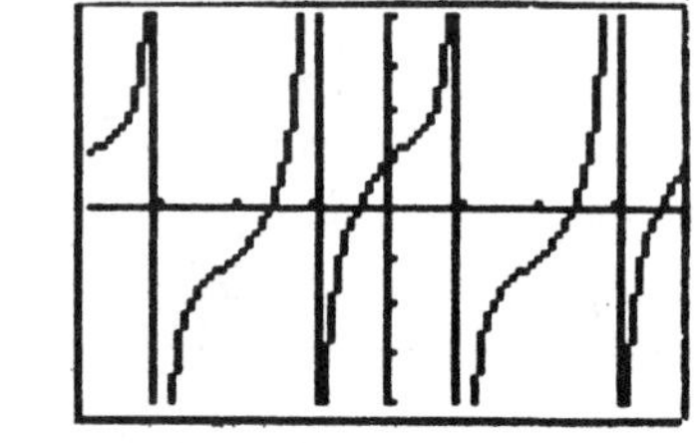

Graph on the TRIG window.

The vertical lines are technology errors.

They occur at the points where $\cos(x) = 0$ and are not in the domain of either function.

Both graphs appear the same so the identity must be proved algebraically.

Use the fundamental Identity on the left.

b.) $(\sin(x))^2 - \cos(x) = (\cos(x))^2 + \sin(x)$

Enter the left side of the equation as Y1, enter the right side of the equation as Y2.

If possible on your calculator, use different displays for each function.

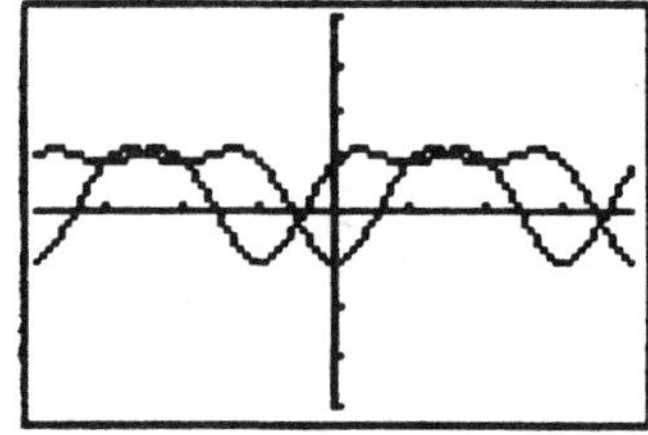

Graph both functions on the TRIG window.

They definitely are not the same graph, so the equation is not an identity.

c.) $\csc(x) - \cot(x) = \frac{\sin(x)}{(1+\cos(x))}$

Enter the left side of the equation as Y1, enter the right side of the equation as Y2.

If possible on your calculator,
use different displays for each function.
Graph both functions on the trig window. There
are points where each function is not defined.
They occur at the points where
$\sin(x) = 0$ or $x = n\pi$ for Y1 and where
$\cos(x) = -1$ or $x = (2n-1)\pi$ for Y2. These
points are not in the domain of both functions.
The graphs look the same, but
the equation is not an identity.

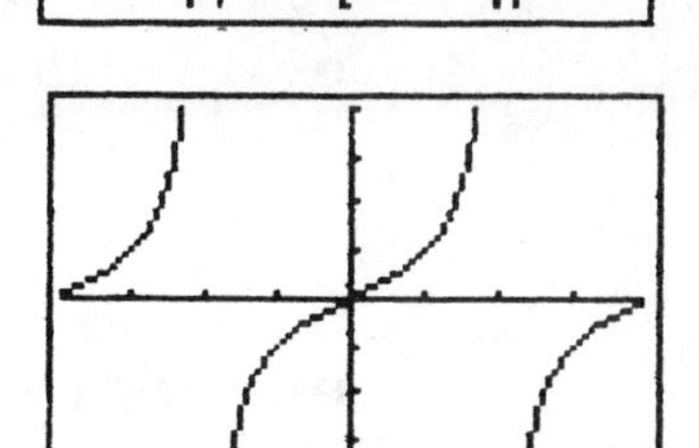

Trigonometric Equations:

Any equation containing trigonometric functions can be solved graphically. It is often easier to use a SOLVER, either GRAPHIC or NUMERIC, after graphing one period of the function. Many SOLVERS need bounds for the roots and a guess to start the solve routine.

Solving Trigonometric Equations:

1. Write the equation in the form $f(x) = 0$.
2. Determine the period p of $f(x)$.
3. Graph $f(x)$ over an interval of length p.
   Find a complete graph.
4. Determine the number of solutions and bounds for each solution.
5. Use a NUMERIC or GRAPHIC root finder to solve the
   equations for the zeros.
6. For each solution, u, on the period graphed,
   all of the solutions to the equation are $u \pm kp$,
   where k is a whole number.

Note:

To return the decimal approximation of special angles in radians back to the exact answer in terms of π, use the identity $x\,rad = \frac{\pi}{n}$ and solve for $n = \left(\frac{x\,rad.}{\pi}\right)^{-1}$. If n is not rational, x is not a special angle.

ex. Write $x \approx .5235987756\,radians$ in terms of π.

$x \div \pi \approx .1666666667$

$ans^{-1} = 6$

$x = \frac{\pi}{6}\,radians.$

ex. Write $x \approx 2.35619449\,radians$ in terms of π

$x \div \pi = .75$

$x = \frac{3\pi}{4}\,radians$

ex. Write $x \approx 1.308996939\,radians$ in terms of π

$x \div \pi \approx .4166666667$

$ans^{-1} = 2.4$

$\rightarrow Frac = \frac{12}{5}$

$x = \frac{5\pi}{12}\,radians$

ex. Write $x \approx .123456789\,radians$ in terms of π

$x \div \pi \approx .0392975165$

$ans^{-1} \approx 25.44690073.$

x is not a special angle.

ex. Solve $sin(2x) = \frac{1}{\sqrt{2}}$ exactly for all solutions on $[0, 2\pi]$

Graph $y = \sin(2x) - \frac{1}{\sqrt{2}}$ on $[0, 2\pi]$ by [-2, 2]

There are 4 roots,

2 in Quadrant 1 and 2 in Quadrant 3.

$0 < x_1 < \frac{\pi}{4}, \frac{\pi}{4} < x_2 < \frac{\pi}{2}, \pi < x_3 < \frac{5\pi}{4}, \frac{5\pi}{4} < x < \frac{3\pi}{2}$

Use the Numeric SOLVER to find each x with the given bounds.

When the solver returns a solution x, the number is stored in the variable x.

Go to the home screen, press x ENTER.

Change the answers to exact form.

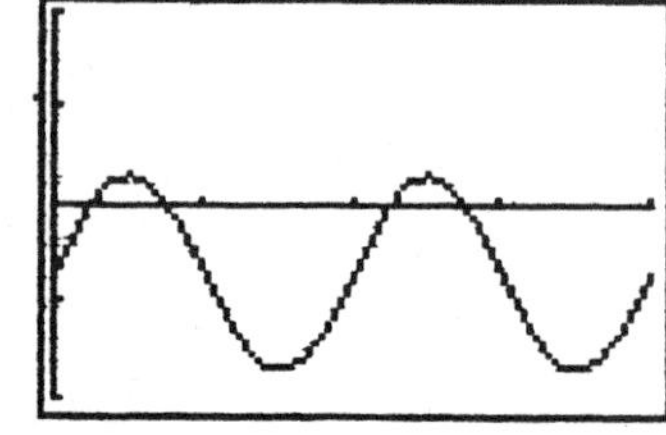

$x_1 \approx .3927... = \frac{\pi}{8}$ use a guess of .5 in the SOLVER.

$x_2 \approx 1.17809... = \frac{3\pi}{8}$ use a guess of 1 in the SOLVER.

$x_3 \approx 3.5342... = \frac{9\pi}{8}$ use a guess of 3 in the SOLVER.

$x_4 \approx 4.3196... = \frac{11\pi}{8}$ use a guess of 4 in the SOLVER.

ex. Find the roots of $3\sin^2(x) - \cos(x) - 2 = 0$ on the interval $0 \le x \le 2\pi$.

A complete graph can be seen on the window $[0, 2\pi]$ by $[-4, 4]$

Both $\sin(x)$ and $\cos(x)$ have period 2π, and the the amplitude of the first term is 3.

There are 4 intersections on this interval, one in each quadrant. Using a NUMERIC SOLVER, you must pick bounds.

If the zeros are called

x_1, x_2, x_3, x_4, then

$0 < x_1 < \frac{\pi}{2}, \; \frac{\pi}{2} < x_2 < \pi,$

$\pi < x_3 < \frac{3\pi}{2}, \; \frac{3\pi}{2} < x_4 < 2\pi.$

These bounds tell the calculator which root to find and a close guess uses less iterations in the SOLVE routine.

Each root must be found separately.

Using a SOLVER, the roots are approximately

$x_1 \approx 1.122,\ x_2 \approx 2.446,\ x_3 \approx 3.837,\ x_4 \approx 5.162.$

To find all of the solutions to the equation,
add $\pm 2k\pi$ (k is a whole number)
to each solution on the given period.

ex. Solve $\sec^2(x) + 5\tan(x) = -2$ for all solutions on $[0, 2\pi]$.

Graph $y = \frac{1}{(\cos x)^2} + 5\tan(x) + 2$ on $[0,2\pi]$ by $[-4, 6]$

There appear to be 4 solutions, however the almost straight vertical lines are technology errors.

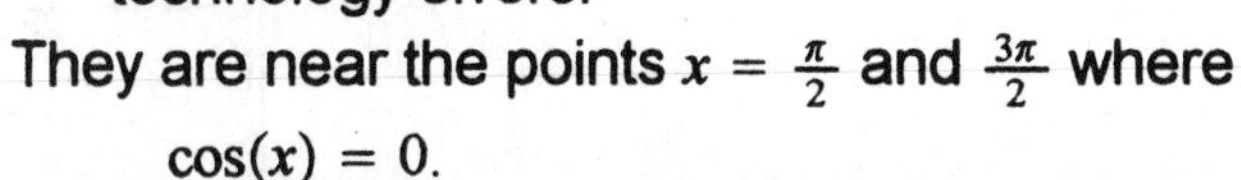

They are near the points $x = \frac{\pi}{2}$ and $\frac{3\pi}{2}$ where $\cos(x) = 0$.

There are 2 solutions, one in quad. 2 and one in quad. 4.

Use a solver to find $x_1 \approx 2.5327...$ and $x_2 \approx 5.6743...$

These angles are not nice multiples of π.

ex. Solve $3\sin(x) = \tan(x)$ exactly for all solutions on $[0, 2\pi]$

Graph $y = 3\sin(x) - \tan(x)$

There are 5 solutions on $[0, 2\pi]$ and 2 technology errors.

The technology errors are near the points where $\cos(x) = 0$.

From the graph it appears that

$x = 0, x = \pi, x = 2\pi$ are solutions.

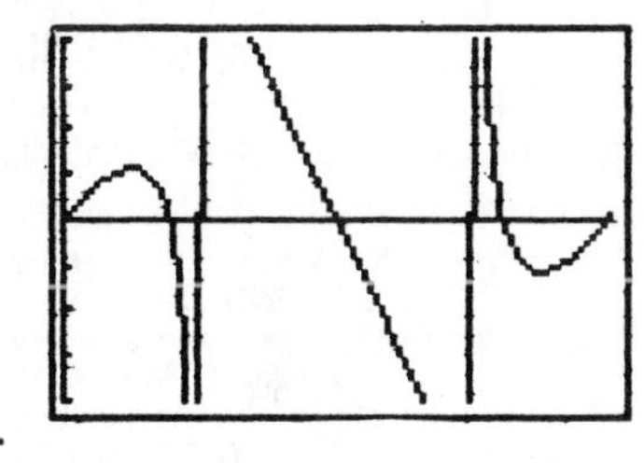

Check them to be sure. $y = 3\sin(0) - \tan(0) = 0$.

$y = 3\sin(\pi) - \tan(\pi) = 0,$

$y = 3\sin(2\pi) - \tan(2\pi) = 0$

The other 2 solutions can be found with a SOLVER.

$x_1 \approx 1.2309...,$

$x_2 = 2\pi - x \approx 5.0522...$

Problems:

Find the solutions to the following equations on $[0, 2\pi]$. *Find exact solutions when possible.*

1. $3\sin(2x) = 1$
2. $\sin(x - 2) = \frac{1}{2}$
3. $\sec(x) + \tan(x) = 3$
4. $3\sin^3(2x) = 2\cos(x)$
5. $\cos(2x) = \cos(x)$

Answers:

1. Graph on [0, 2π] by [-6, 4]

   $x_1 \approx .1699...,$

   $x_2 = \frac{\pi}{2} - x_1 \approx 1.4008...,$

   $x_3 = \pi + x_1 \approx 3.3115...,$

   $x_4 = \frac{3\pi}{2} - x_1 \approx 4.5424...$

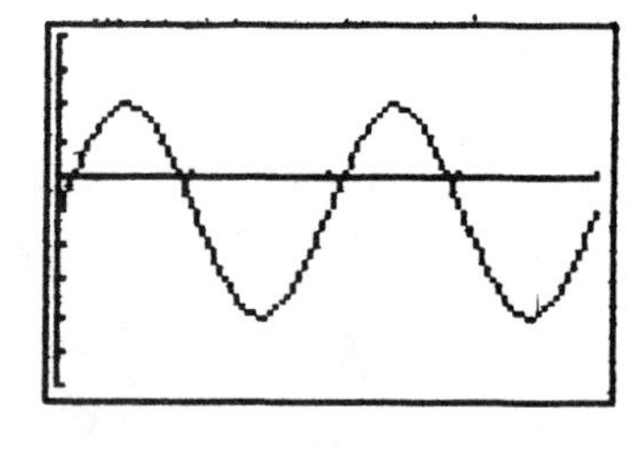

2. Graph on [0, 2π] by [-2, 2]

   $x_1 \approx 2.523...,$

   $x_2 \approx 4.6179...$

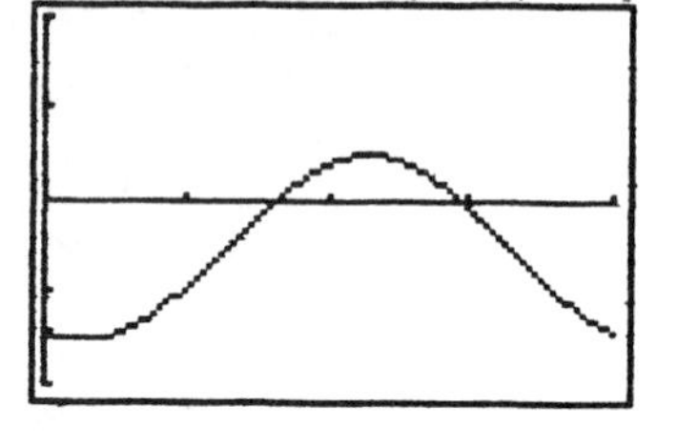

3. Graph on [0, 2π] by [-5, 5]

   $x \approx .9272...$

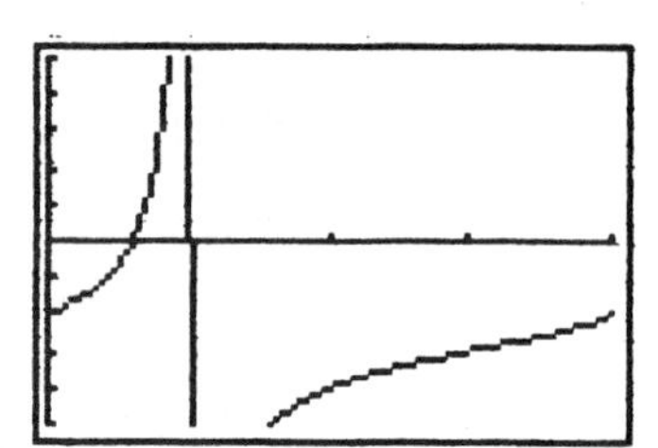

4. Graph on [0, 2π] by [-4, 4]

   $x_1 \approx .4958...,$

   $x_2 \approx 1.2538...,$

   $x_3 = \frac{\pi}{2} \approx 1.5708...$

   $x_4 \approx 1.8877...,$

   $x_5 \approx 2.6457...,$

   $x_6 \approx 4.7123...$

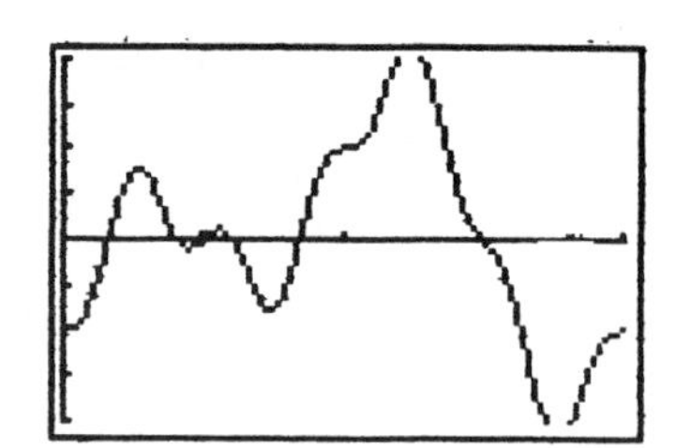

5. Graph on [0, 2π] by [-2, 2]

   $x_1 = \frac{2\pi}{3} \approx 2.0943...,$

   $x_2 = 2\pi - x_1 = \frac{4\pi}{3} \approx 4.1887...$

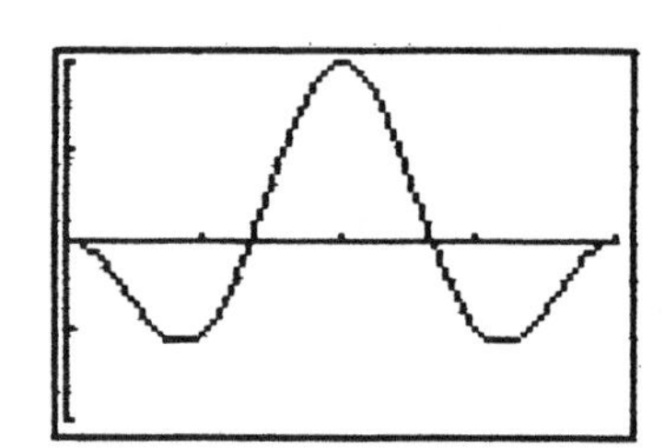

Inverse Trigonometric Functions:

Solve $\sin(x) = .5$ for x. Graph $y = \sin(x)$ and $y = .5$ on the default trig window.
There are an infinite number of angles x that satisfy both equations.
We define one of them, the angle between $-\pi$ and π to be the inverse function.
Use $x = \sin^{-1}(y)$ as the notation to solve for x.
Now interchange x and y.
The inverse function is

$$y = \sin^{-1}(x) \text{ for } -\pi \le x \le \pi.$$

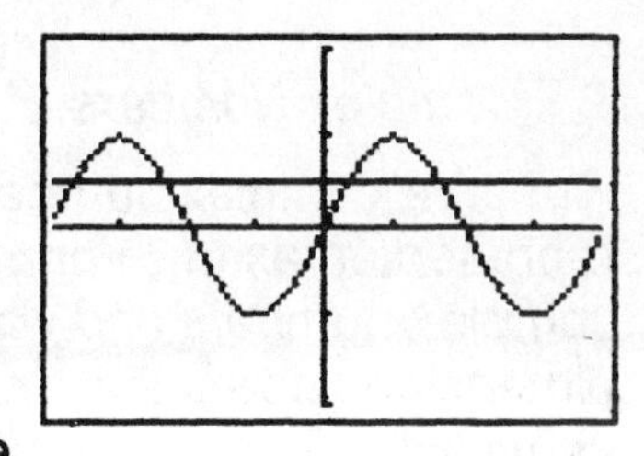

This key $\sin^{-1}$ is the shift key
for sin on your calculator.

The calculator only handles functions,only one
solution can be found using the inverse key.

More solutions can be found by
using the symmetries of the graph or
by using either the graphic or Numeric SOLVER.

The inverse keys work well when solving triangles where
the solution is an angle less than 90°.

A SOLVER is more useful when solving an equation
with more than one solution.

Applications of Trigonometry

Complex Numbers:

A complex number is an ordered pair of real numbers. A complex number can be represented as (a,b) or $a+bi$. Consult directions for your calculator to find out how to enter complex numbers and what calculations it can do. Complex numbers can be plotted in 2 dimensions called the *Complex Plane*. The horizontal axis is called the *Real* axis and the vertical axis is called the *Imaginary axis*.

The modulus of a complex number is the distance from the origin to the complex number in the complex plane.

Use the absolute value function.

ex. $|3+2i| = abs(3+2i) = \sqrt{13}$

Every complex number can be written in polar form $r(\cos\theta + i\sin\theta) = re^{i\theta}$.

Some calculators will change complex numbers between rectangular and polar form.

Selections in the MODE menu determine if the complex numbers will be displayed in complex rectangular or complex polar form.

Use the usual keys for addition, subtraction, multiplication and powers.

ex. $(1+2i)+(3-5i) = (4-3i)$

$(1+2i)\times(3-5i) = 13+i \approx 13.038e^{.077\theta}$

$(1+i)^3 = -2+2i = \sqrt{2}e^{\frac{i\pi}{4}} = \sqrt{2}(\cos(\frac{\pi}{4}) + i\sin(\frac{\pi}{4}))$

Complex roots can be found graphically. See EXAMPLE 6. in the text.

Find the fifth roots of unity.

Set the MODE menu to Parametric.

Set the window to [-1.5,1.5] by [-1, 1] and

$0 \le t \le 2\pi, \quad t-step = \frac{2\pi}{5} \quad (2\pi \div no.of roots)$

Use [-1.7, 1.7] by [-1,1] on wide screen calculators to plot the circle so that it is its true shape.

Enter $x1 = \cos(t) \quad y1 = \sin(t)$ and graph.

The calculator plotted only 5 points and connected them with straight lines.

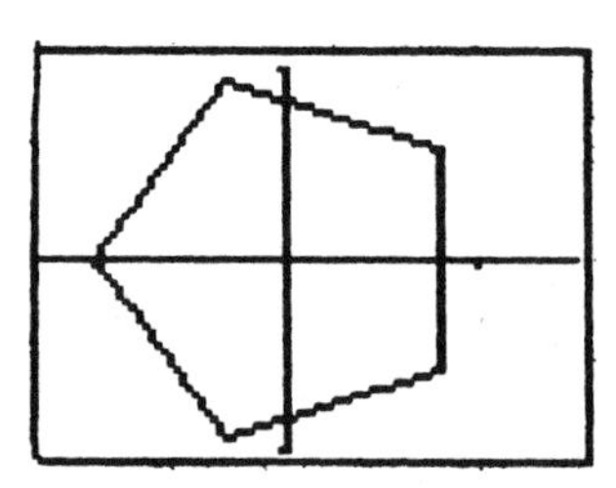

Trace to read the values, $(0,0), (.309,.951)$ $(-.809,.951), (-.809,-.588), (.309,.951)$.

These values are $\cos\frac{2k\pi}{5} + i\sin\frac{2k\pi}{5} \quad k = 0,1,2,3,4$

The angle in radians can be changed to a multiple of π by the method in Chap. 8.

The angles are $t = 0, \frac{2\pi}{5}, \frac{4\pi}{5}, \frac{6\pi}{5}, \frac{8\pi}{5}$

Vectors in the Plane:

Calculators consider vectors as a special case of Matrices that have either 1 row or 1 column. Use [ ] to represent a vector.

To enter a vector use the matrix editor or enter them on the home screen.

Every vector $\overrightarrow{PQ}$ from the point P to the point Q is equivalent to a vector $\overrightarrow{OR}$ with initial point at the origin.

If $P = (x_1,y_1)$ and $Q = (x_2,y_2)$ then

$\overrightarrow{PQ} = \overrightarrow{OR}$ where $R = (x_2 - x_1, y_2 - y_1)$

On a calculator this vector is written as a row matrix $[\,[x_2 - x_1, y_2 - y_1]\,]$, or as a column matrix $[\,[x_2 - x_1]\,, [y_2 - y_1]\,]$,

Vector Arithmetic:

Vector addition, multiplication by a constant and the dot product are done with matrices.

ex. Let $[A] = [1,2]$, $[B] = [3,-4]$

$[A] + [B] = [4,-2]$

$-2[A] = [-2,-4]$

the dot product $[a] \bullet [B]$

$$= [1,2]\begin{bmatrix} 3 \\ -4 \end{bmatrix} = [1 \times 3 + 2 \times -4] = -5$$

```
[A]+[B]
              [[4  -2]]
-2[A]
             [[-2  -4]]
[A]*[B]ᵀ
                [[-5]]
■
```

The length (magnitude) of a vector

$\|A\| = \sqrt{1^2 + 2^2} = \sqrt{5}$

The direction of a vector is a vector of length 1 in the same direction.

The direction of $[B]$ is $\frac{[3,4]}{5}$.

Analytic Geometry

Parametric Graphs:

Some curves in space can not be represented by a function $y = f(x)$.
However by bringing in a third variable t, it is possible to
represent the curve with 2 functions $x = f(t)$ and $y = g(t)$.
Now more information is available, the path and the location at
time t.
This new variable t is called a parameter and the equations defining
x and y are called parametric equations.
Parametric Equations are not unique.
Many sets of parametric equations can represent any given curve.

To Graph Parametric equations:

First change function to parametric in the MODE or SETUP MENU.
In the range WINDOW, set the Xrange, the Yrange and the TRange.
The Tstep determines how much t changes each time a point is plotted.
TRACE works in Parametric MODE. If the graph is on a Friendly Window,
the answers still depend on the parameter t.

ex. Show how changing t and $t - step$ changes a parametric graph.
Graph $x = t^3 - 10t^2 + 29t - 10$ and
$y = t^2 - 7t + 14$, $0 \le t \le 6.5$ $tstep$ .2
on $[-12, 35]$ $scl5$ by $[-5, 20]$ $scl2$
using Friendly Window $k = 5$.

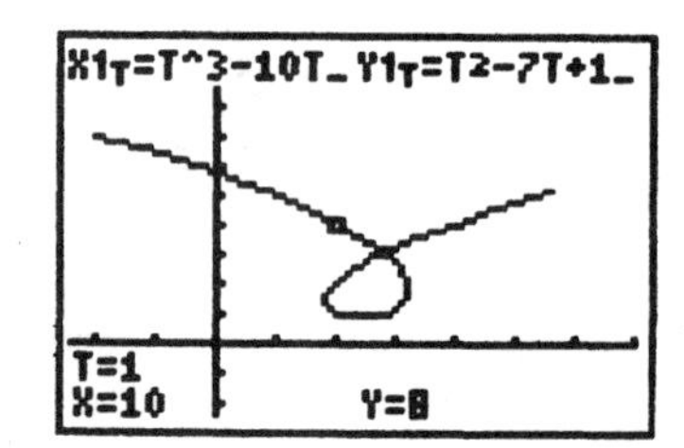

Trace to see the x and y values
for $t = 0, 1, 2, 3, 4, 5, 6$
Repeat changing the $tstep$ to $.5, .8, 1, 1.5$
Trace on each of these windows
to see which points are plotted.

The change in $tstep$ changes how often a point is graphed.
Now change the range for t, letting Tmax $= 5, 6, 7$
Tmin and Tmax determine where (x and y values)
the curve starts and stops.

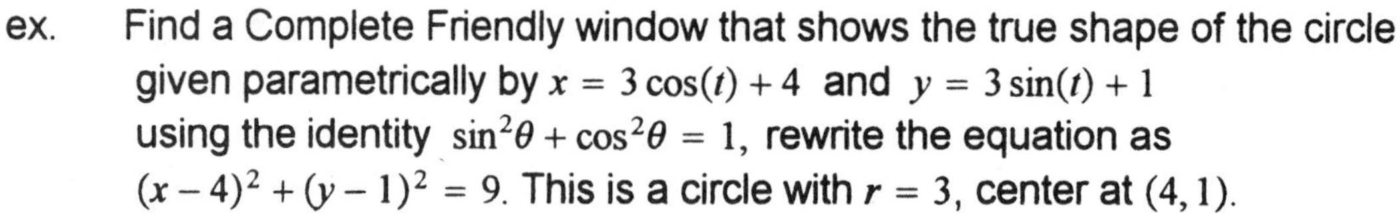
ex. Find a Complete Friendly window that shows the true shape of the circle
given parametrically by $x = 3\cos(t) + 4$ and $y = 3\sin(t) + 1$
using the identity $\sin^2\theta + \cos^2\theta = 1$, rewrite the equation as
$(x - 4)^2 + (y - 1)^2 = 9$. This is a circle with $r = 3$, center at $(4, 1)$.
Set the T range to $[0, 2\pi]$ with $t\ step = \frac{\pi}{24}$.
Graph on the Decimal Window.

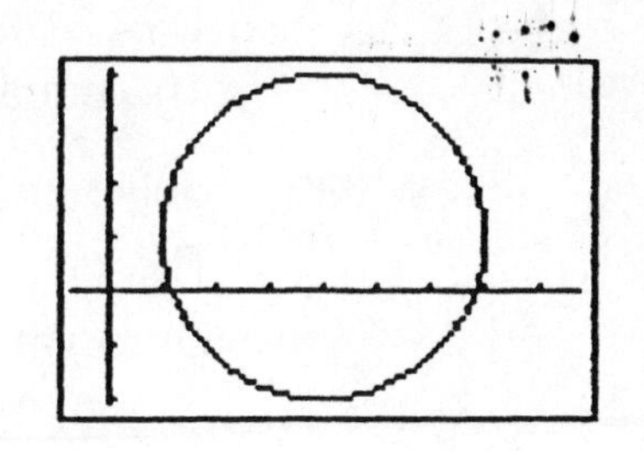

Add the required amount to the numbers
in Xrange and Yrange to see the entire circle.
Use the built in window SQUARE if the
circle is not round.
On a 94 by 62 window, add 4 to the
numbers in the X range,
1 to the numbers in the Y Range.
Trace. Use the left arrow to trace around
the curve in the direction of increasing t.

Conic Sections:

When a right circular cone is cut by a plane, the intersection is a curve called a *conic section*. The non-degenerate conics are the circle, ellipse, hyperbola and parabola. The degenerate forms are a point, a straight line and 2 intersecting straight lines.

The general equation for the conic sections is

$$ax^2 + bxy + cy^2 + dx + ey + f = 0$$

If $b = 0$, then the conic is not rotated,
the major axis is parallel to the x or y axis.
If $d = 0$ and $e = 0$ the conic is in standard position
and the center is at the origin.

Graphing Conics:

The important thing in graphing conics is to graph on a Friendly window where there is a Complete graph.

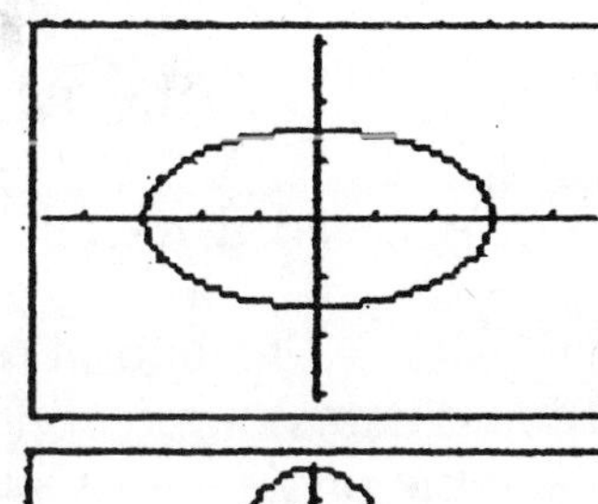

ex. 1 Graph $4x^2 + 9y^2 = 36$

Solve for $y = \frac{\pm\sqrt{36-4x^2}}{3}$

Graph on the decimal Window.
On most calculators both equations
can be graphed at the same time.
Trace to find important points.

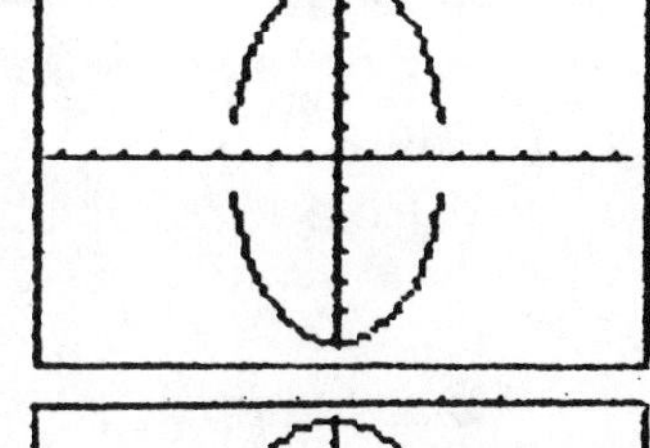

ex. 2 The equation to graph is $3x^2 + y^2 = 36$

Solving: $y = \pm\sqrt{36 - 3x^2}$

Graph on the decimal window, then zoom out
with the zoom factors both set at 2.

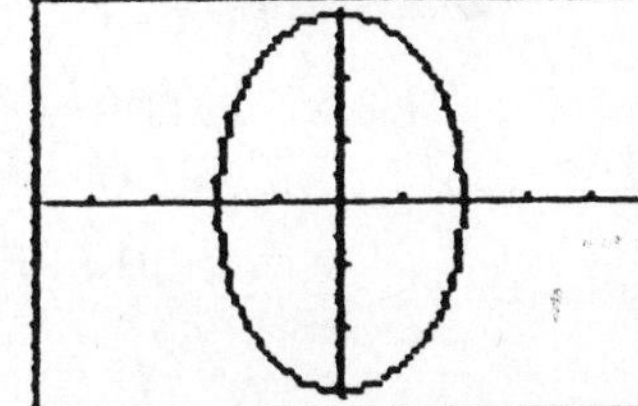

ex. 3 The equation to graph is $9x^2 + 4y^2 = 36$

Solving: $y = \pm\frac{1}{2}\sqrt{36 - 9x^2}$

Graph on the decimal window.
Trace to see the important points.

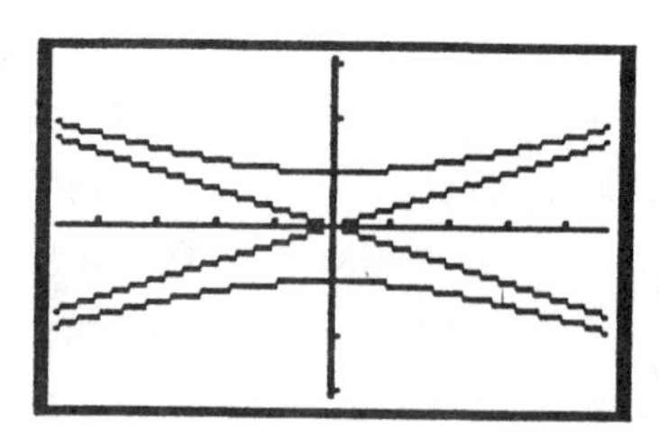

ex. 4. The equation to graph is $9y^2 - x^2 = 9$

The equation is a hyperbola.

Solving: $y = \pm\frac{\sqrt{9+x^2}}{3}$

The equations of the asymptotes are $y = \pm\frac{x}{3}$

Graph the equation and its asymptotes on the Decimal Window.

Trace to find the important points.

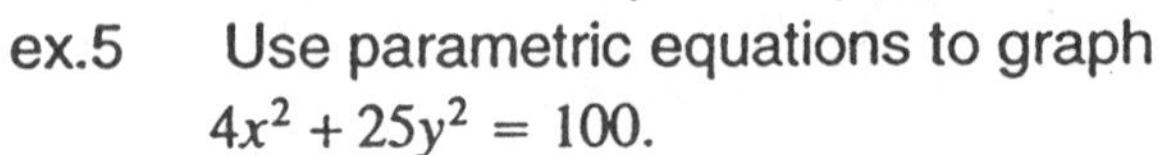

ex.5 Use parametric equations to graph $4x^2 + 25y^2 = 100$.

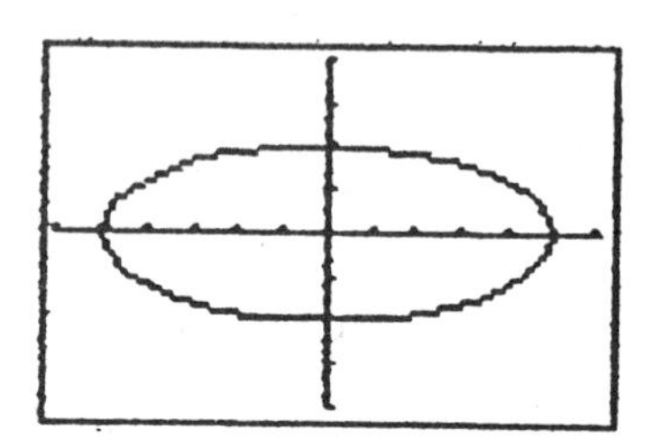

Change the equation to parametric form

$x = 5\cos(t) \quad y = 2\sin(t)$ is one possibility.

If the entire ellipse does not show on the decimal window, use [-6,6] by [-4, 4] and $0 \le t \le 2\pi$, $tstep = \frac{\pi}{24}$

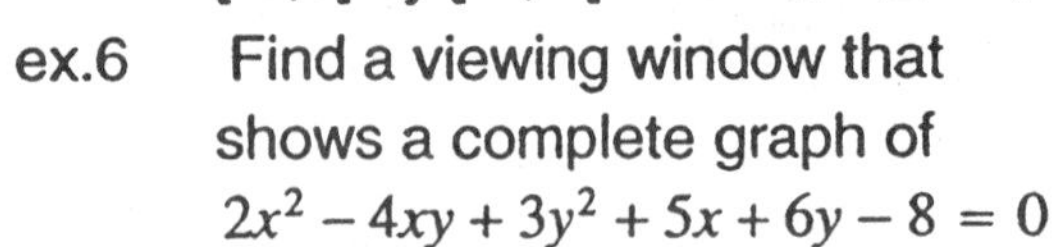

ex.6 Find a viewing window that shows a complete graph of

$2x^2 - 4xy + 3y^2 + 5x + 6y - 8 = 0$

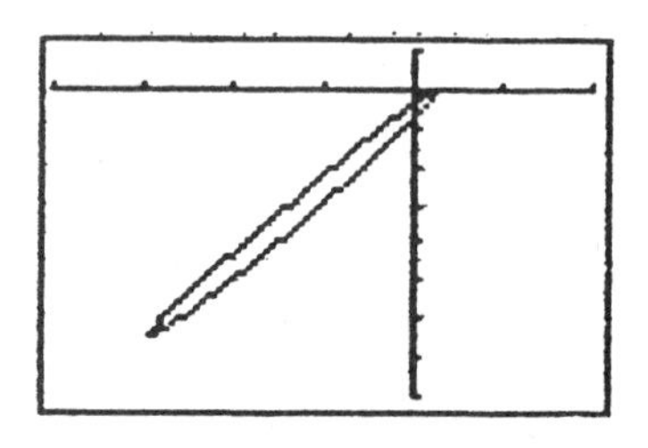

The discriminant $B^2 - 4AC = -8$.

The equation is an ellipse.

Solve $y = \frac{-(-4x+6) \pm \sqrt{(-4x+6)^2 - 12(2x^2+5x-8)}}{6}$

Graph on the Standard Window.

Change the window to [-20,10] by [-80,10]

The graph is not its true shape.

The major axis is along the line $y = 4x - 6$

Polar Coordinates:

Instead of representing a point in space by its x and y coordinates, (rectangular coordinates), represent the point by a line from the origin to the point, called the radius r, and the angle the radius makes with the positive x axis called θ.

Conversions between Polar and Rectangular conversions can be done on most calculators.

Polar Coordinate representations are not unique.

Polar Graphs:

To set Polar Mode on the calculator, change to Polar in the MODE or SET UP Menu. The $y =$ menu will now read $r = \theta$ and the variable key will now print θ.

Polar equations are just a special case of parametric equations.
If $r = f(\theta)$, then $x = r\cos(\theta) = f(\theta)\cos(\theta)$ and $y = r\sin(\theta) = f(\theta)\sin(\theta)$

ex. Find a complete graph of $r = 1 + \sin(\theta)$

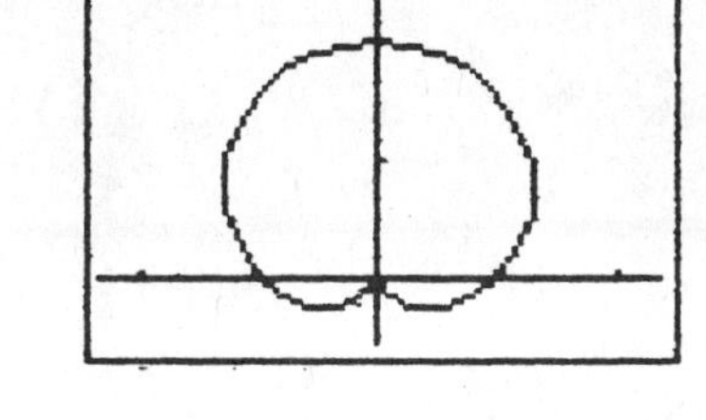

Set the window to $0 \le \theta \le 2\pi$,
then graph on the decimal window.
This is a complete graph, but
the graph is small.
Zoom in by a factor of .5, then
move the origin down
1 unit on the Yaxis,
by adding 1 to Ymin and Ymax.

ex. Find a complete graph of $r = 2\sin(2\theta)$

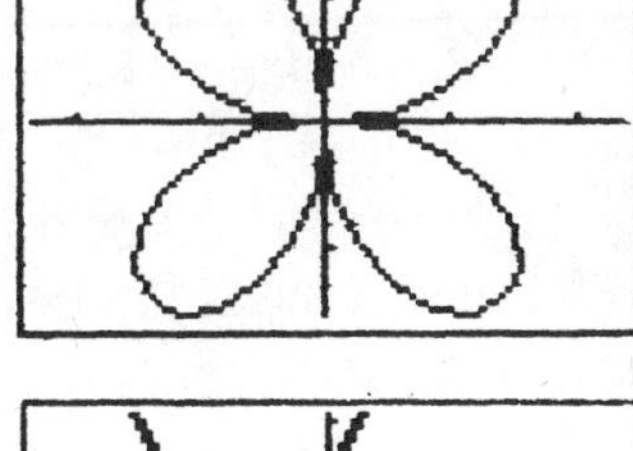

Set the window to $0 \le \theta \le 2\pi$, then
graph on the decimal window.
This is a complete graph, but
the graph is small.
Zoom in by a factor of .5.

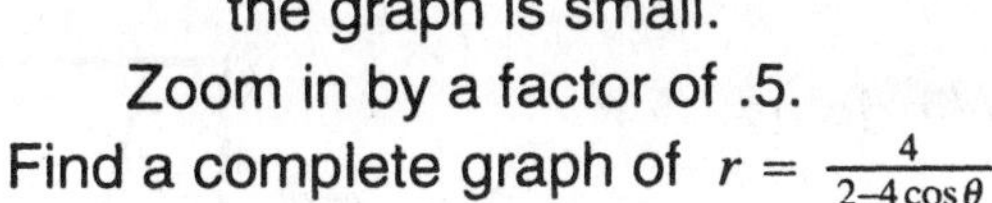

ex. Find a complete graph of $r = \frac{4}{2-4\cos\theta}$

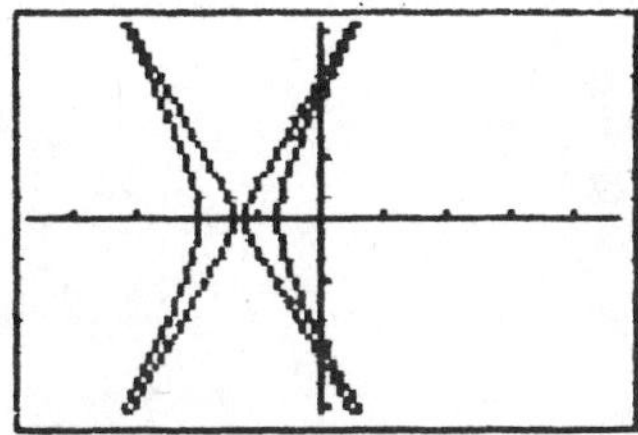

Set the window to $0 \le \theta \le 2\pi$, then
graph on the decimal window.
This is the complete graph of a hyperbola.
Note the false asymptotes that are
really a technology error.

Problems:

Find complete graphs for the following polar functions.

1. $r = 3\cos(4\theta)$
2. $r = \frac{2.1}{1+.7\cos\theta}$
3. $r = \frac{8}{1-\cos\theta}$
4. $r = \frac{10}{4-3\sin\theta}$

Answers:

1. Decimal Window

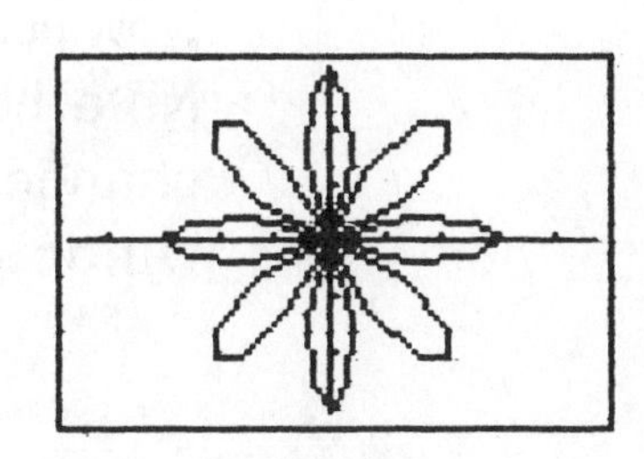

2. [–7.7, 1.7] by [–3.1, 3.1]

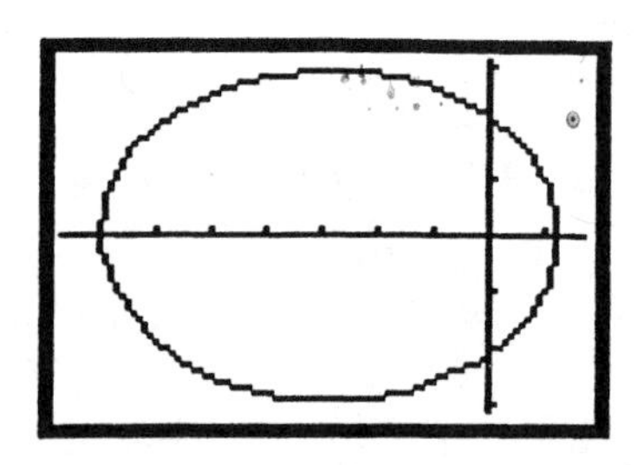

3. Decimal Window
Zoom Out factors = 4

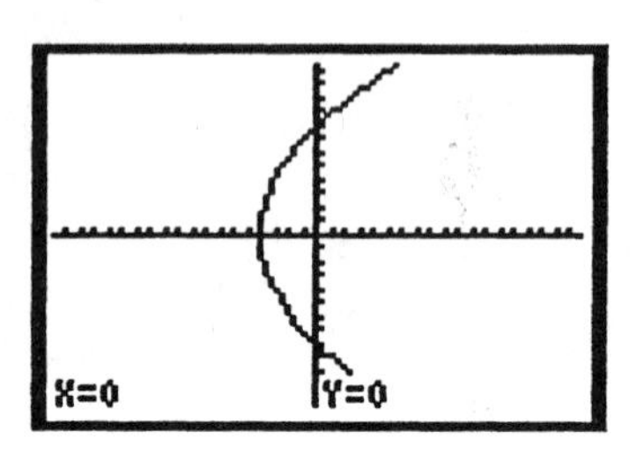

4. Decimal Window

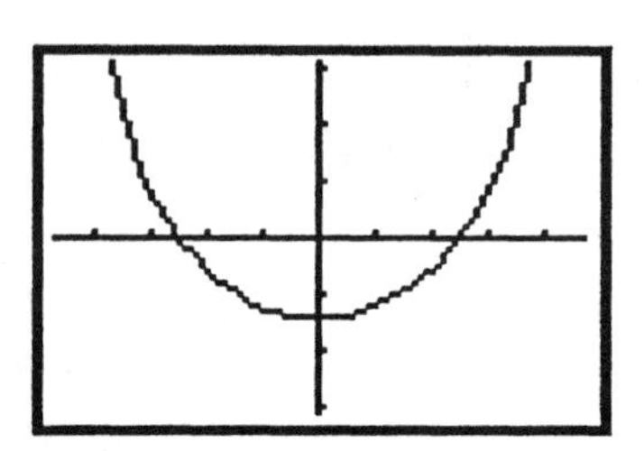

Multiply Yrange ×2, then add 4.

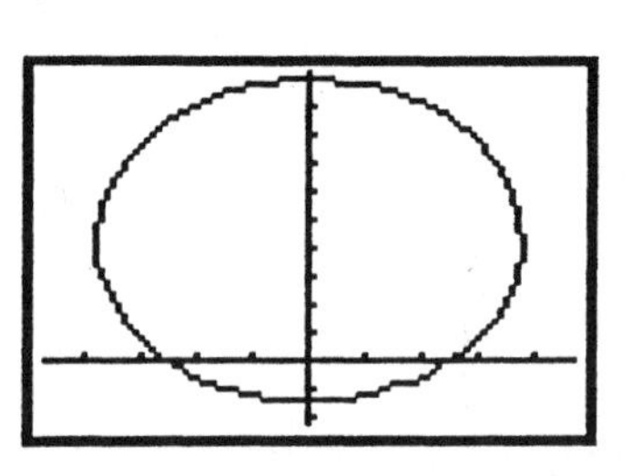

Now use SQUARE.
Note how the graph in exercise 4 changes shape in the different windows.

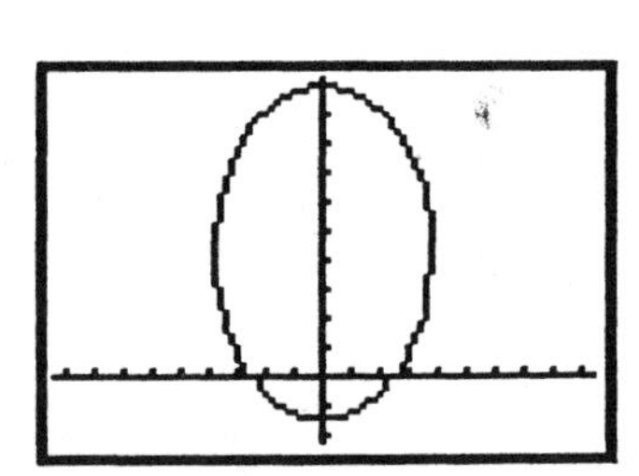

Systems of Equations

Systems of Linear Equations:

A solution of a system is a set of values for the variables that satisfies all of the equations.

Graphical Solutions of Linear Systems with 2 variables:

ex. Linear systems in 2 variables can be solved graphically.

Solve the system on a graphing calculator.

$2x - y = 1$

$3x + 2y = 4$

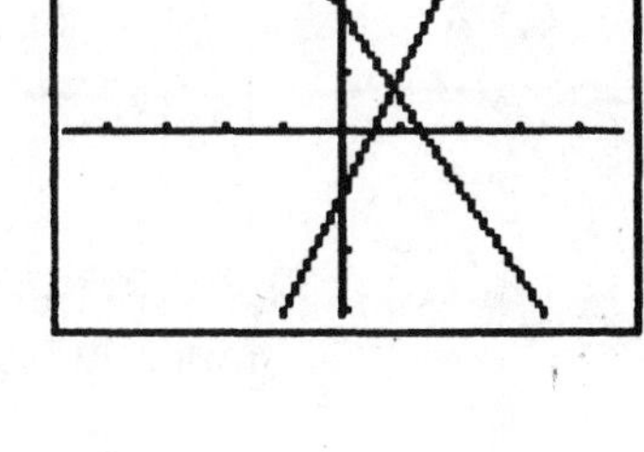

Solve the equations for y and enter them into the calculator.

$y = 2x - 1$

$y = (-3x + 4)/2$

Graph these on the decimal window.

Trace to the intersection.

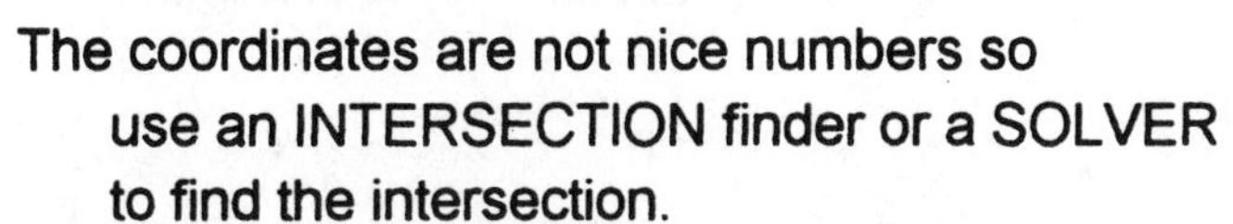

The coordinates are not nice numbers so use an INTERSECTION finder or a SOLVER to find the intersection.

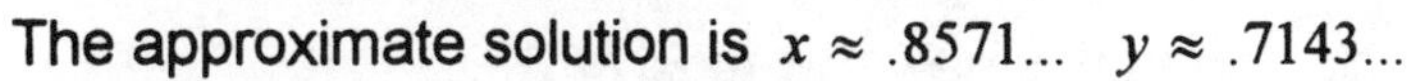

The approximate solution is $x \approx .8571...$ $y \approx .7143...$

These values, to calculator accuracy, are stored in x and y.

To check the answers to this problem, go to the home screen.

Press x enter, then y enter. The correct values are displayed.

Enter the left side of one of the equations. $2x - y$ enter.

The result is <u>1</u>.

Check the other equation in the same way, using the same values for x and y. It checks also.

ex. Solve the system

$2x - 3y = 5$

$x - 1.5y = .25$

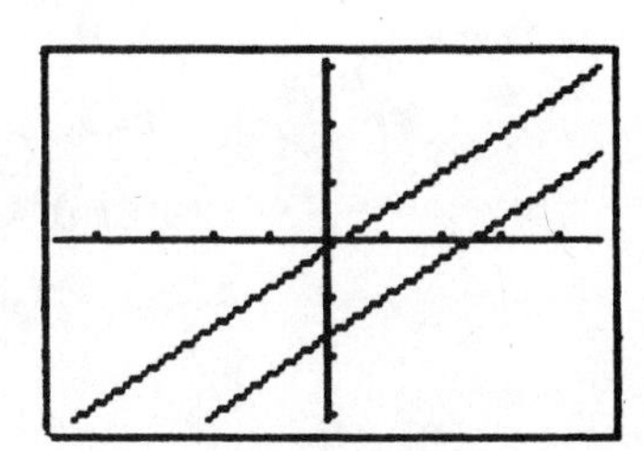

Solving both equations for y,

$y = \frac{2}{3}x - \frac{5}{3}$ and $y = \frac{4}{6}x - \frac{1}{6}$

Graph these equations, the lines are parallel, so there is no solution.

Solving Linear Systems with Row Operations

The commands to perform basic matrix row operations are in the MATRIX MATH or OPERATIONS menu. They are:

1. Interchange rows
2. Multiply a row by a constant
3. Multiply a row by a constant and add it to another row.

ex. Solve the system:

$$\begin{aligned} x + 4y - 3z &= 1 \\ -3x - 6y + z &= 3 \\ 2x + 11y - z &= 0 \end{aligned}$$

First enter the augmented matrix into the calculator

$$\begin{bmatrix} 1 & 4 & -3 & 1 \\ -3 & -6 & 1 & 3 \\ 2 & 11 & -1 & 0 \end{bmatrix}$$

Store the matrix as matrix [A]

Multiply Row 1 by 3 add this new row to Row 2

$$\begin{bmatrix} 1 & 4 & -3 & 1 \\ 0 & 6 & -8 & 6 \\ 2 & 11 & -5 & 0 \end{bmatrix}$$

Multiply Row 1 by -2 add this new row to Row 3

Store the matrix as matrix [B]

$$\begin{bmatrix} 1 & 4 & -3 & 1 \\ 0 & 6 & -8 & 6 \\ 0 & 3 & 1 & -2 \end{bmatrix}$$

Multiply Row 3 by -2 add this new row to Row 2

Store the matrix as matrix [B]

$$\begin{bmatrix} 1 & 4 & -3 & 1 \\ 0 & 0 & -10 & 10 \\ 0 & 3 & 1 & -2 \end{bmatrix}$$

Interchange Row 2 and Row 3

Store the matrix as matrix [B]

$$\begin{bmatrix} 1 & 4 & -3 & 1 \\ 0 & 3 & 1 & -2 \\ 0 & 0 & -10 & 10 \end{bmatrix}$$

Solve this new system by back solving:

$-10z = 10 \rightarrow z = -1$

$3y - z = -2 \rightarrow 3y - (-1) = -2 \rightarrow y = \frac{-1}{3}$

$x + 4y - 3z = 1 \rightarrow x + 4\left(\frac{-1}{3}\right) - 3(-1) = 1 \rightarrow x = \frac{-2}{3}$

Solving Linear Systems with Gauss-Jordan Method:

Solving systems using row-reductions (Gaussian elimination) and back substitutions can be done on a graphing calculator.

However, it is much more efficient to use the Gauss-Jordan Method.

ex. Solve the system:

$$x + 4y - 3z = 1$$
$$-3x - 6y + z = 3$$
$$2x + 11y - z = 0$$

First enter the augmented matrix into the calculator

$$\begin{bmatrix} 1 & 4 & -3 & 1 \\ -3 & -6 & 1 & 3 \\ 2 & 11 & -1 & 0 \end{bmatrix}$$

Store the matrix as matrix [A]

From the MATRIX MATH or OPS menu,

enter *rref*(*matrixA*) using the notation for matrix A that your calculator uses, enter.

Change the decimals to fractions.

The following matrix is displayed.

$$\begin{bmatrix} 1 & 0 & 0 & \frac{-2}{3} \\ 0 & 1 & 0 & \frac{-1}{3} \\ 0 & 0 & 1 & -1 \end{bmatrix}$$

This augmented matrix is equivalent to the system:

$$x = \frac{-2}{3}, \quad y = \frac{1}{3}, \quad z = -1$$

The Gauss-Jordan Method will solve the system, even if the solution is not unique.

ex. Solve the system

$$-2x + 3y - 2z = 16$$
$$-5x + 3y - 5z = 22$$
$$x + \qquad z = -2.$$

The augmented matrix for this system is:

$$\begin{bmatrix} -2 & 3 & -2 & 16 \\ -5 & 3 & -5 & 22 \\ 1 & 0 & 1 & -2 \end{bmatrix}$$

The reduced row echelon form is

$$\begin{bmatrix} 1 & 0 & 1 & -2 \\ 0 & 1 & 0 & 4 \\ 0 & 0 & 0 & 0 \end{bmatrix}$$

This matrix is equivalent to the set of equations:

$x + z = -2, \quad y = 4.$

For each choice of z there is a different set of points
that satisfies all of the equations.

Choose z to be anything, call $z = \alpha$.

Then $x = -2 - \alpha, \quad y = 4, \quad z = \alpha$ is a solution for the system.

As there are many choices for α, the system has an infinite number of solutions.

Other Matrix Methods:

Using Cramers Rule or Finding the inverse to solve the matrix equation **AX**=**B** work only if the system has a unique solution.

With some real data, the algorithms used to find inverses
and determinants will give incorrect answers,
often because of round off errors.

Matrix Arithmetic:

To add subtract or multiply matrices, use the usual +,−,× keys.

The matrices **must** be entered onto the home screen
from the matrix names in the matrix menu.

Division is done by multiplying by the inverse.

The inverse is computed with the x^{-1} key.

The identity matrix of any order is found in the matrix MATH menu.

Non-linear Systems:

Graphic solutions to non-linear systems are very practical
whenever both equations can be solved for y.

ex. Solve the system:

$x^2 + y^2 = 8$

$x^2 - y = 6$

Solve both equations for y.

$Y1 = \pm\sqrt{8 - x^2} \qquad Y2 = 6 - x^2$

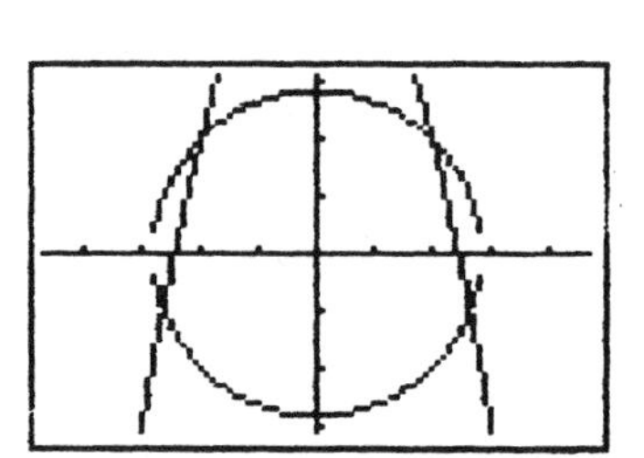

Graph these equations on the Decimal Window.

There are 4 intersections, one in each quadrant.

It will be easier to find the zeros,
rather than the intersections.

Graph $Y1 - Y2$ there are 2 curves, $+\sqrt{8 - x^2} - Y2$

and $-\sqrt{8-x^2} - Y2$

Trace to the smaller positive zero.

If the window is Friendly, read $x = 2$

Trace to the larger positive zero.

This intersection is not "nice",

use a SOLVER $x \approx 2.6457...$

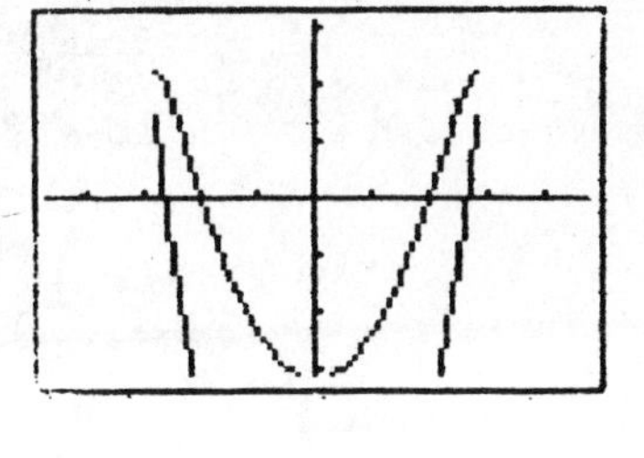

As both curves are symmetric to the Y axis, the other values are the negatives of these.

To find the y values, substitute into $y = 6 - x^2$.

The solutions are:

$x = \pm 2,\ y = -2 \qquad x \approx \pm 2.6457...,\ y = 1$

(Note) as y is an integer, $x = \pm\sqrt{7}$

To see this after solving for the largest x, go to the home screen, and enter x. The decimal equivalent to the accuracy of the calculator is displayed.

Press x^2, the integer 7 is displayed.

ex, Solve the system:

$$2x + y = 4$$
$$xy = 2$$

Solve both equations for y

$$y = 4 - 2x \qquad y = \tfrac{2}{x}$$

Graph on the decimal window.

The only intercepts are in quadrant 1

Change the window :

Change Xmin and Ymin to 0.

Graph then TRACE. It appears that the line might be tangent to the curve.

Trace to $x = 1, y = 2$. Use the [△] and [▽] keys.

The Trace coordinates do not change, so this point lies on both graphs.

Trace to points where both graphs appear the same.

There is no other point that lies on both of the graphs.

Using the Intersection Solver confirms this result.

Problems:

Solve the following systems:

1. $y = x^2 - 2x + 5$
   $y = x + 9$
2. $x^2 + y^2 = 4$
   $y = x^2 - 1$

3. $y = e^x$
   $xy = 1$
4. $y = e^{2x} - 2$
   $2x - 5y = 3$
5. $4x^2 - 6xy + 2y^2 - 3x + 10y = 6$
   $4x^2 + y^2 = 64$

Answers:

1. Decimal Window
   Plot $Y1 - Y2$
   $x = -1, 4$

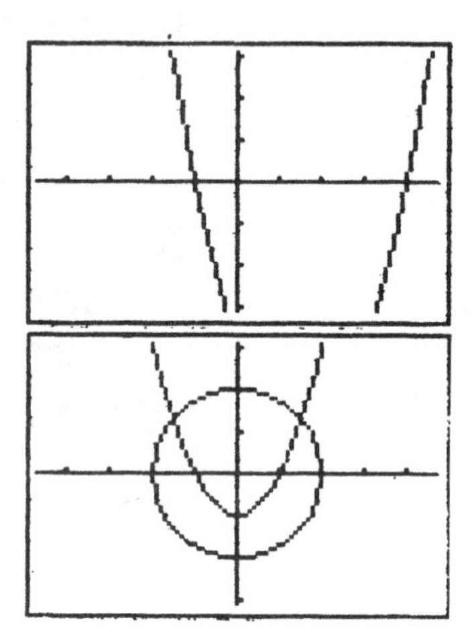

2. Decimal Window
   Plot $Y1$ and $Y2$
   Both intersections are on
   $Y1 = +\sqrt{}$ so delete $Y1 = -\sqrt{}$
   Use a Graph Solve
   $x \approx \pm 1.517...,\ y \approx 1.303...$

3. Decimal Window
   Plot $Y1$ and $Y2$
   $x \approx .5671...$
   $y \approx 1.763...$

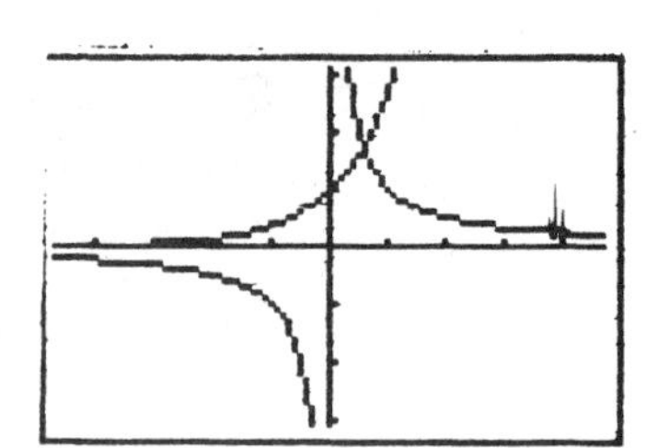

4. $[-4.7, 4.7]$ by $[-10.2, 10.2]$
   Plot $Y1$ and $Y2$
   $x \approx .1954...,\ y \approx -.5218...$
   $x \approx -3.498...,\ y \approx -1.999..$

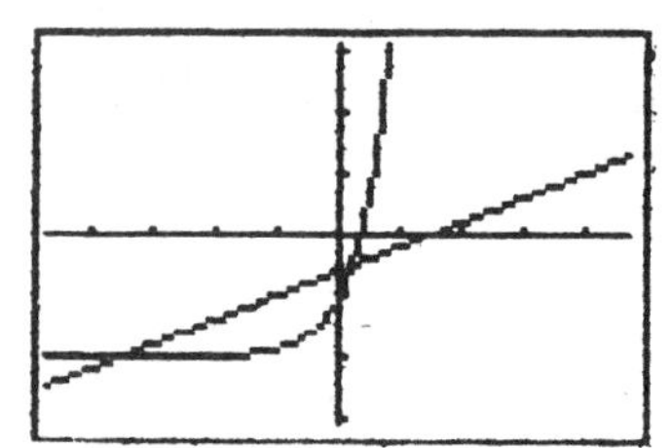

5. $[-4.7, 4.7]$ by $[-10.2, 10.2]$
   Solve both equations for $Y1$ and $Y2$
   Graph. Both intersections are on
   $Y2 = -\sqrt{}$ Delete $Y2 = +\sqrt{}$.
   Graph the left branch of $Y1$ with $Y2$.
   Use a SOLVER, $x \approx -.2047...,\ y \approx -7.989...$
   Graph the right branch of $Y1$ with $Y2$.
   Use a SOLVER, $x \approx .7287...,\ y \approx -7.866...$

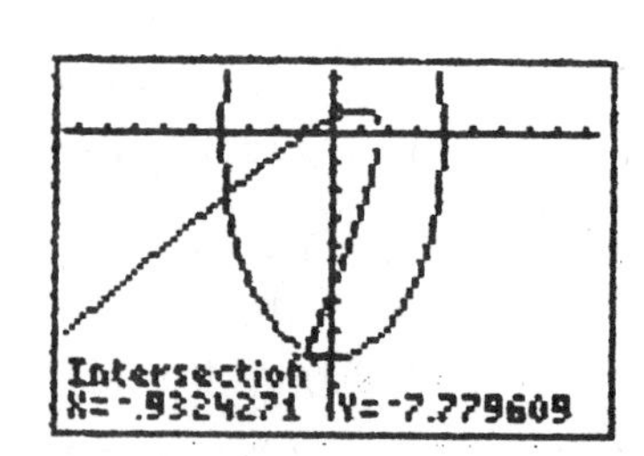

Discrete Algebra

Sequences are ordered lists of numbers where there is a rule or pattern to find the number in the sequence.

Graphing calculators will write out a finite number of terms in a sequence when a formula for the nth or general term of the sequence is given. Often all of the terms will not fit on to the screen use the right arrow to see the rest of them.

Sequences may also be view in a TABLE.

A sequence is defined recursively, if the first term is given and the formula computes the next term from the current term. The Casio calculators have a special recursive menu.

For a finite sequence the partial sums and the sum of the sequence can be found by the calculator.

ex. $\Sigma_{k=1}^{8} \frac{1}{k^3}$

a) Enter the sequence and store it as **L1**.
   seq($1 \div k^3, k, 1, 8, 1$) store in **L1**.

```
seq(1/K^3,K,1,8,
1)
{1 .125 .037037...
Ans→L1
{1 .125 .037037...
Ans►Frac
{1 1/8 1/27 1/6...
```

b) Find the partial sums.
   CumSum(**L1**)

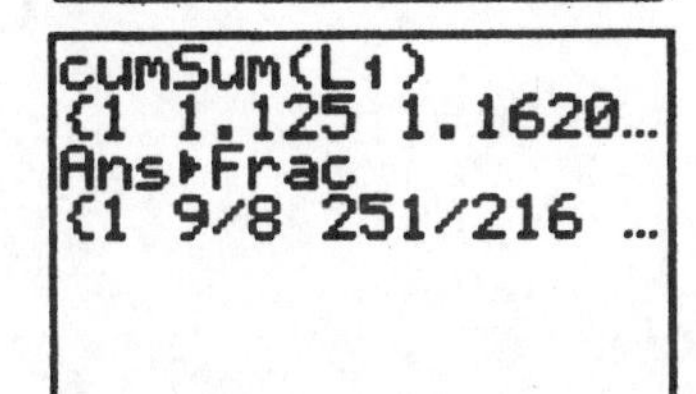

c) Find the sum of the sequence.
   Sum(**L1**)

```
sum(L1)
          1.195160244
```

Arithmetic Sequence:

An arithmetic sequence is one where the difference of any consecutive terms is a constant.

This sequence can be generated recursively

$$a_n = a_{n-1} + d$$

or with a general term.

$$a_n = a_1 + (n-1)d$$

where a_n is the general or last term,

a_1 is the first term

n is the number of terms

d is the common difference.

Problem:

1. $\Sigma_{k=1}^{25}(\frac{k}{4}+5)$

   a) Generate the sequence and store it in **L2**.

   b) Find the sequence of partial sums for **L2**.
   Find the sum of the terms in **L2**.

Answers:

a)

```
seq(K/4+5,K,1,25
,1)
{5.25 5.5 5.75 ...
Ans►Frac
{21/4 11/2 23/4...
Ans→L2
{5.25 5.5 5.75 ...
```

b)

```
cumSum(L2)
{5.25 10.75 16....
Ans►Frac
{21/4 43/4 33/2...
sum(L2)
              206.25
```

Geometric Sequence:

A geometric sequence is one where the quotient of any consecutive terms is a constant.

This sequence can be generated recursively

$a_n = ra_{n-1}$

or with a general term.

$a_n = a_1 + (n-1)d$

where a_n is the general or last term,

a_1 is the first term

n is the number of terms

r is the common ratio

Problem:

1. $\Sigma_{k=1}^{9} 3(\frac{-1}{2})^k$.
   a) Generate the sequence and store it in **L3**.
   b) Find the sequence of partial sums for **L3**.
   c) Find the sum of the terms in **L3**.

Answers:

a)

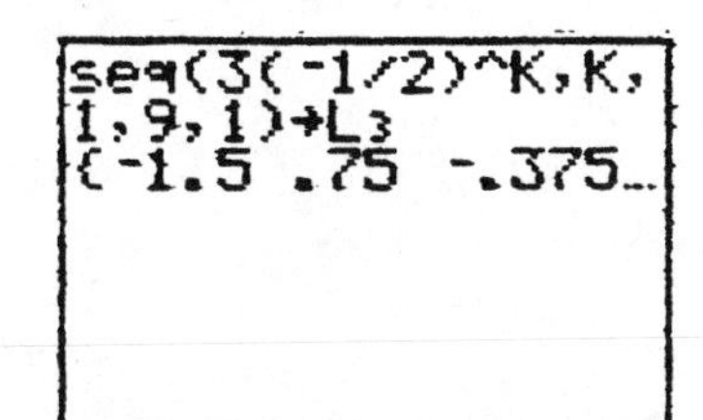

b)

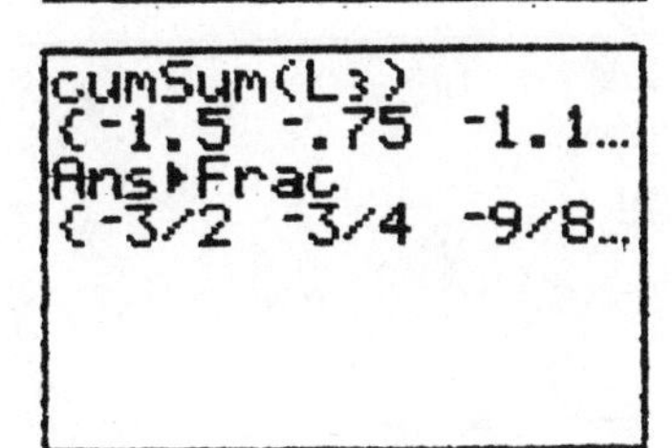

c)

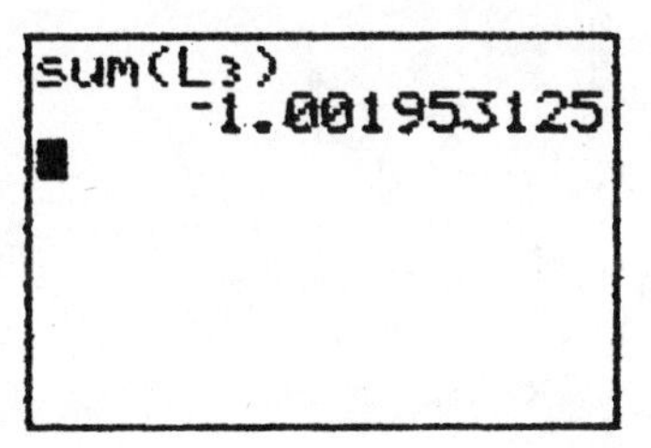

Binomial Expansion:

$(x+y)^n = x^n +$

The binomial coefficients $_{n0}C_x$ are found in the Probability menu in the calculator.

These coefficients can be found using a Table.

Chapter 3 – Using the TI-83/TI-84 Plus Graphing Calculator.

1. Notation:

Keystrokes, except for numbers are in bold brackets **[MATH]**

For a second function $\sqrt{\ }$, press **[2nd]** [$\sqrt{\ }$] the second functions are printed on the left over the keys.

For the letter A, press **[ALPHA][A]** the ALPHA symbols are printed on the right over the keys.

For functions in the screen menus 2^3, press **2 [MATH] 3** or press **2 [MATH]** then arrow down to highlight [**3**: ^{3}] and press **[ENTER]**
In this text 2^3 is keystroked **2 [MATH][3**: ^{3}]

Calculator results are underlined.

2. Basics:

To adjust the screen contrast: Press **[2nd]** then [△] to increase the contrast or **[2nd]** then [▽] to decrease the contrast.

Mode settings: Press **[MODE]** select all of the settings at the left side of the screen menu. Leave these settings unless directed to change a mode setting.

3. Arithmetic Computations:

Order of Operations: The order is Algebraic, operations are performed from left to right. First exponentiation, then multiplication and division, then addition and subtraction. Parenthesis must be used to change the algebraic order. Parenthesis must be used around the numerator and denominator in Algebraic Fractions.

Minus sign: Use [(–)] for negative, [–] for subtraction.

Fractions: to enter fractions use [÷]

To change a decimal to a fraction use **[MATH][1:▸Frac][ENTER]**

To change a fraction to a decimal use **[MATH][2:▸Dec][ENTER]**

To calculate: press **[ENTER]**

ex. $2 \times 3 + 4 \times 5$ **[ENTER]** 26

$2 \times (3 + 4) \times 5$ **[ENTER]** 70

$3 + 4 \div 2$ **[ENTER]** 5

$(3 + 4) \div 2$ **[ENTER]** 3.5

[MATH][1:▸Frac][ENTER] 7/2

4. **Scientific Notation**:

From the keyboard use number between 1 and 10 **[2nd][EE]** exponent.
The exponent must be a number between -99 and 99.
To Display answers in Scientific Notation change the MODE menu entry.
[MODE][▷][ENTER][CLEAR]
You may want to change the number of decimal places displayed.
To display 3 places, change to FIX 3 press
[MODE][▽][▷][▷][▷][▷][ENTER][CLEAR]
To return to Normal:
[MODE] Highlight Normal **[ENTER][▽]** highlight Float **[ENTER][CLEAR]**

5. **Editing**:

*Before typing [**ENTER**]*
A blinking box determines the current position on the screen.
The arrow keys move the cursor around the screen.
To change an entry move the blinking cursor to the entry, then
type the new entry. It will replace the old entry.
To delete a symbol, move to the symbol, press **[DEL]**
To insert a symbol, move to the symbol after the insertion point,
press **[2nd][INS]** then type the new text.
*After typing [**ENTER**]*
Press **[2nd][ENTRY]** the last command is recalled, edit it
as explained above in *Before typing ENTER*
You may continue to press **[2nd][ENTRY]** to go back to other commands.
To clear the home screen press **[CLEAR]**
To exit a MENU, press **[2nd][QUIT]** or enter another MENU.
To delete data in MEMORY, press **[2nd][MEM][2:Delete]** then delete
from any of the given MENUS. Press **[2nd][QUIT]** when done.
To erase a function from the Y = screen, highlight any symbol in the function,
press **[CLEAR]**

6. **Entering Algebraic Functions**:

From the Y = screen, use the arrow keys to move to an empty position.
Enter the function. It is now available to use as a graph or a table. This menu holds 10 functions.
On the home screen, enter the function. When you press **[ENTER]** it will be
evaluated with the values of the variables that are stored in Memory.

The Catalog is a list of all functions, operations and symbols in the calculator.

Press **[2nd][CATALOG]** you will see a list sorted alphabetically.
Press the first letter of the word you want, use the down arrow to move the pointer to the entry, press **[ENTER]** and the word is pasted onto the home screen.

The symbols are listed after Z. From A, use the up arrow to display the symbols.

When you enter built in functions, left parenthesis are included. Complete the input variable, then type the right parenthesis.

The key **[x,T,θ,n]** prints x in Function mode, T in parametric mode, θ in polar mode and n in sequential mode.

Special Functions:

Absolute value of x: press **[MATH][▷][1:abs][x,T,θ,n]**

Powers, x^5: press **[x,T,θ,n][^] 5**

2^x: press **2 [^][x,T,θ,n]**

e^x: press **[2nd][e^x][x,T,θ,n]** the ^ is forced.

Roots, $\sqrt[5]{x}$: press **5 [MATH] [5:$\sqrt[x]{\ }$][x,T,θ,n]**or **[x,T,θ,n][^][(]1 [÷] 5 [)]**

Natural log, $ln(x+5)$: press **[ln][x,T,θ,n][+] 5 [)]**

Conjugate Pairs, $\pm\sqrt{4-x^2}$: press **[{] 1,-1 [}][2nd][$\sqrt{\ }$] 4 [–] [x,T,θ,n][x²][)]**

Factorial, 5!: press **5 [MATH][▷][▷][▷][4:!]**

Trig function, $tan(3x^2)$: press **[tan] 3 [x,T,θ,n][x²][)]**

Inverse trig function, $arcsine(x)$: press **[2nd][sin⁻¹][x,T,θ,n][)]**

Combinations, ${}_5C_2$: press **5 [MATH][▷][▷][▷][3:${}_nC_2$] 2**

7. Evaluating a function: (checking your answer)

Variables are represented as single alphabetic characters, A through Z, and θ. Numbers are stored in these positions using the command **[STO▸]** an arrow → is shown on the home screen.

When an algebraic expression is entered, the variables are replaced with the stored constants and a number is displayed.

To evaluate a function on the home screen use the key **[ALPHA]** [:]to connect commands.

ex. To evaluate $x^2 + 3x - 1$ for $x = 2$,
Press **2 [STO▸][x,T,θ,n][ALPHA][:]**
[x,T,θ,n][x²][+] 3 [x,T,θ,n][–] 1
[ENTER]
the result is 9.

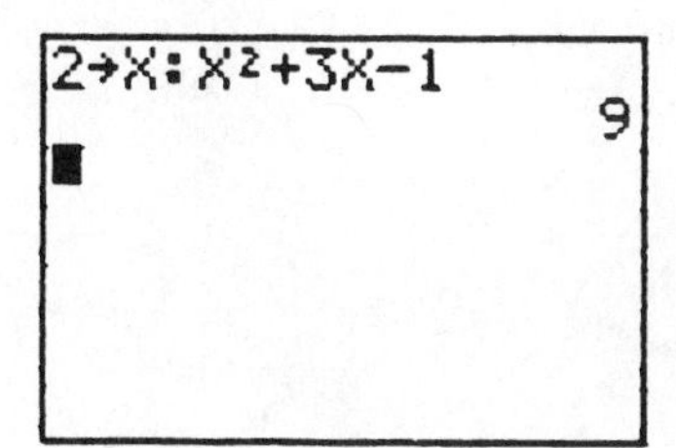

If the function is stored in the Y = menu it may be recalled from the VARS menu.

ex. Enter the function $x^2 + 3x - 1$ in the Y = menu as Y1.
Press the blue key [Y=] press [**CLEAR**] then
[**x,T,θ,n**][**x²**][+] **3** [**x,T,θ,n**][–] **1** [**ENTER**][**2nd**][**QUIT**]
Store the number 2 in memory x , **2** [**STO▸**][**x,T,θ,n**][**ENTER**]
Press [**VARS**][▹][**1:Function**][**1:Y1**][**ENTER**]
the result is 9.

All expressions containing an x will be evaluated with $x = 2$ until this number stored in the x box is replaced with another number.

8. Angles:

The type of measure for angles is set in the MODE menu.
Set [**MODE**] to either **Radian** or **Degree** by highlighting the entry and pressing [**ENTER**]
After the MODE is set, use the operations in [**2nd**][**ANGLE**] to change the units.

To enter an angle in Degrees while in **Radian** MODE:
ex. Press [**SIN**] **30** [**2nd**][**ANGLE**][**1:°**][)][**ENTER**] **.5**

To enter an angle in Radians while in **Degree** MODE:
ex. Press [**SIN**] $\pi \div$ **6** [**2nd**][**ANGLE**][**3:ʳ**][)][**ENTER**] **.5**

To change Rectangular Coordinates to Polar (Degree Mode):
ex. Press [**2nd**][**ANGLE**][**5:R▸Pr(**] **1,1** [)][**ENTER**]
1.4142... is the radius in polar form.
Press [**2nd**][**ANGLE**][**6:R▸Pθ(**] **1,1** [)][**ENTER**]..
45 is the angle in **Degrees.**

To change Polar Coordinates (Degree Mode) to Rectangular:
ex. Press [**2nd**][**ANGLE**][**7:P▸Rx(**] **$\sqrt{2}$,45** [)][**ENTER**] 1
is the x coordinate in rectangular coordinates.
Press [**2nd**][**ANGLE**][**8:P▸Ry(**] **$\sqrt{2}$,45** [)][**ENTER**] 1
If the MODE is Radians, enter the angle in Radians.

9. Building Tables:

The TI-83/TI-84 has a built in TABLE function used to evaluate functions for different values of x. The algebraic expression to be evaluated is stored in the Y = menu. In the TABLE SET menu, set the initial value for x, and the step size between entries, or set ASK to enter a random set of numbers for x.

ex. Enter $x^2 + 3x - 1$ in Y1 in the Y = Menu.
Press **[2nd][TBLSET]** set the following values:
[(–)] 4 [ENTER] for TblStart
1 [ENTER] for ΔTbl
Highlight **Auto** for both Indpnt and Depend
Press **[2nd][QUIT]** when finished.
Press **[2nd][TABLE]** to view the table.

X	Y1	
-4	3	
-3	-1	
-2	-3	
-1	-3	
0	-1	
1	3	
2	9	

X= -4

Use up and down arrows to see more values in the table.
Tables may also be constructed in the STAT menu to examine and evaluate statistical data.

10. Building Lists:

Numbers can be stored in Lists either on the home screen or in the STAT EDIT menu.
Lists may be named (stored) using the **[2nd]** keys **[L1]**,**[L2]**,**[L3]**,**[L4]**,**[L5]**,**[L6]** or by naming them.

ex. Enter the numbers 1, 3, 5, 7 in list **L1**
[2nd][{] 1 , 3 , 5 , 7 [2nd][}][STO▸][2nd][L1][ENTER]

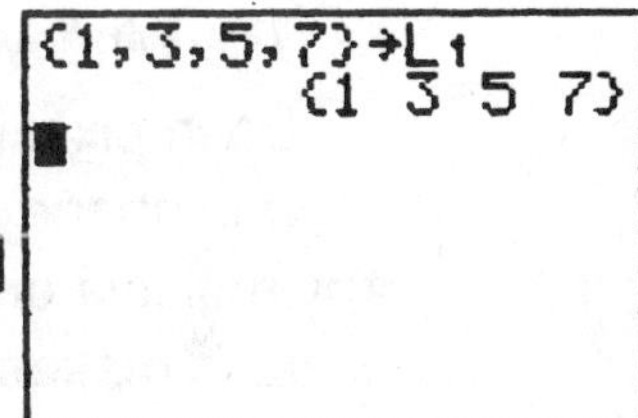

Lists from the keyboard are stored in the STAT EDIT menu.
To display L1 on the home screen, press **[L1][ENTER]**
Arithmetic operations are done using the +,–,×,÷,^ keys,
Other operations are done from the **[2nd][LIST]** **OPS** and MATH menus.

Special Lists:

OPS [**5**:**seq**(] expression, variable, begin, end increment)
Generate a sequence of numbers using a rule for expression.

ex. **[2nd][LIST][▷] OPS**
[5:seq(][x,T,θ,n][x²] , [x,T,θ,n] , 1 , 5 , 2][)][ENTER]
square the numbers starting with 1, adding 2, ending at 5.
The result is { 1, 9, 25}

OPS [**6**:**cumSum**(] (list) returns a list of cumulative sums starting with the first entry. For L1, { 1 , 4 , 9 , 16 }

MATH [5:**sum**(] (list) returns the sum of all the numbers in the list. For L1, the sum is 16.

OPS [7:△**List**(] (list) returns a list of the differences of consecutive terms. For L1, the list is { 2 , 2 , 2 }
This list is 1 shorter than the original list.

11. Graphing Functions:

The function to be graphed must be entered as one of the 10 functions

In the Y = menu. Enter $x^2 + 3x - 1$ in Y1.

The function must be turned on to graph, so,
the = must be highlighted.
To turn the function on or off,
highlight the = and press [**ENTER**]
Each ENTER reverses the current state.

```
Plot1 Plot2 Plot3
\Y1=X²+3X-1
\Y2=
\Y3=
\Y4=
\Y5=
\Y6=
\Y7=
```

To graph the function press the blue key [**GRAPH**]
The function is graphed on the current window settings.

The ZOOM menu has choices to zoom in or out
and for built in window sizes.

Press the blue key [**ZOOM**] press [**4**:**ZDecimal**]
The window is set to a Friendly window
Each pixel is 0.1 units, and the graph is its
true shape. Unit size on X axis = unit size on Y axis.

Any multiple or translation of these settings
also makes a Friendly window.

Press [**ZOOM**][**6**:**ZStandard**] This is the Standard window, it sets
X to [-10,10] and Y to [-10,10]. The units on the X and Y axis are not the same size, so graphs are not their true shape.

Press [**ZOOM**][**5**:**ZSquare**] Now the graph is its true shape again.
The window settings for the Y axis are left the same, and the X settings have been recalculated so that the units on the X and Y axis are the same. This is called a Square window. It usually is not Friendly.

Press [**ZOOM**][**0**:**ZoomFit**] This is the AutoScale function.
After the Xmax, Xmin are set, it finds a Y-range to show a
Complete Graph. The Y settings often need adjustment to
get a clear picture. With ZoomFit a Friendly Window stays Friendly.

Press the blue key [**WINDOW**] It displays the current settings for the X and Y ranges. These values can be changed by using the arrow keys to highlight the entry, pressing [**CLEAR**] then typing the new value.

Use **[ENTER]** after each change.
(Remember to use (–) for a negative number).
The current values can be multiplied by a constant, or a constant can be added to an entry and the calculator will do the arithmetic.
Press **[GRAPH]** to return to the graph.

Press the blue key **[TRACE]** A blinking star cursor appears on the Y axis.
The left and right arrow keys move this cursor along the function graph.
The coordinates that are displayed are points that satisfy the function equation. The Y values are computed from the X value of the pixel.
When more than one function is plotted, the up and down arrows move the cursor vertically between the different plots. These coordinates are only approximations unless the graph is on a Friendly window and the coordinate is a rational number.

To graph a _split function_: $f(x) = \left\{ \begin{array}{ll} x+7, & x \le -5 \\ 4-x, & x > -5 \end{array} \right\}$

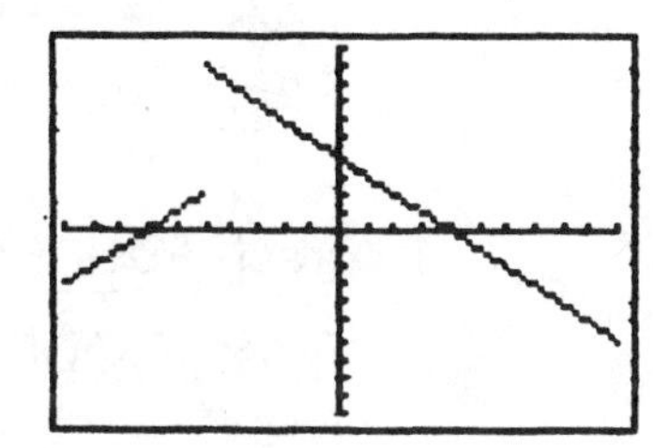

Press **[Y=]** then enter the function in 2 empty spaces.
$Y1 = (x+7) \div (x \le -5) \qquad Y2 = (4-x) \div (x > -5)$
$Y1$ = [(][**x,T,θ,n**][+] **7** [)][÷]
[(][**x,T,θ,n**][**2nd**][**TEST**][**6**:≤][(–)] **5** [)]
$Y2$ = [(] **4** [–][**x,T,θ,n**][)][÷][(][**x,T,θ,n**]
[**2nd**][**TEST**][**3**:>][(–)] **5** [)]
Graph using **[ZOOM][6:ZStandard]**

To graph _Parametric Equations_:

Press **[MODE]** arrow down to Func,
Use [▷] to highlight Par, then press **[ENTER]**
Press **[Y=]** the menu now reads $\mathbf{X_{1T}}$= and $\mathbf{Y_{1T}}$=
You may want to change the window settings.
Now X and Y are both functions of the independent variable T.
The settings for Tmin and Tmax determine how much of the graph is plotted.
Tstep determines how many points are plotted.
Enter the functions and press **[GRAPH]**

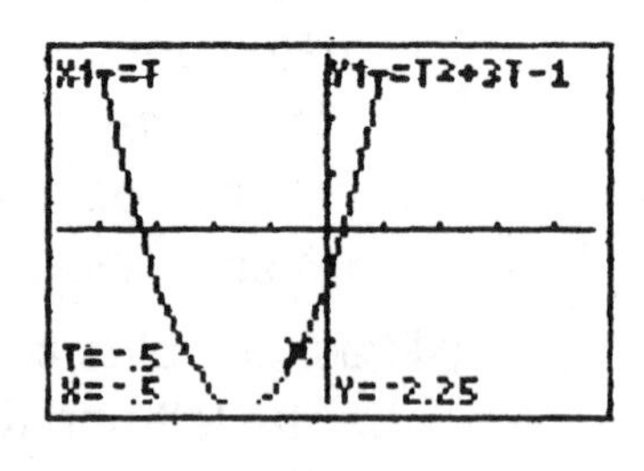

ex. Write the function $y = x^2 + 3x - 1$
in parametric form.
Let $x = t, \quad y = t^2 + 3t - 1$.
Enter $\mathbf{X_{1T}} = \mathbf{T}$
$\mathbf{Y_{1T}} = \mathbf{T^2+3T - 1}$
Press **[WINDOW]**
Set **Tmin** = [(-)] **5**, **Tmax** = **5**.
Press **[ZOOM] 4** The graph is displayed.
Use **[TRACE]** This is **not** a Friendly Window

To graph *implicit functions*: easily,

Write the equation in Parametric form.

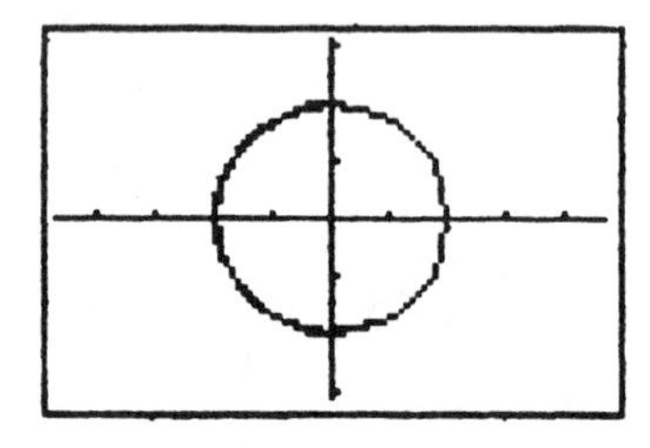

ex. Write the equation $x^2 + y^2 = 4$
in parametric form: $x = 2\cos(t), \quad y = 2\sin(t)$.
Enter $\mathbf{X_{1T}} = \mathbf{2}\cos(\mathbf{T}), \mathbf{Y_{1T}} = \mathbf{2}\sin(\mathbf{T})$
[ZOOM] 4 The graph is a circle.

Remember to change back to Function Mode.

To graph *Polar Equations*:

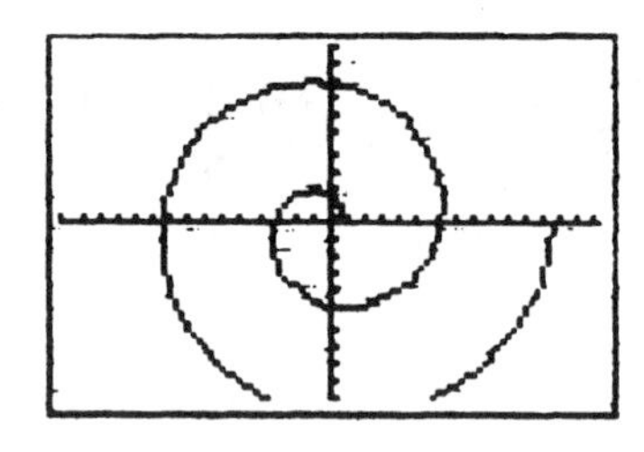

Press **[MODE]**, arrow down to FUNC
Use [▷][▷] to highlight Pol, then press **[ENTER]**
Press **[Y=]** Enter **r1** = θ, where r = f(θ).
Polar is a special case of Parametric.
Press **[WINDOW]** Set $\theta\text{min} = 0$, $\theta\text{max} = 4\pi$
θstep = $\frac{\pi}{24}$ $[-10, 10]$ by $[-10, 10]$ **[GRAPH]**
[ZOOM][5:ZSquare] for the true shape.

Remember to change back to Function Mode.

12. Solving Equations:

Write the equation in the form $f(x) = 0$
Using Trace and Zoom will give you an approximate answer.
Use either a Zoom box or use Zoom In.

To Set the ZOOM Factors:

Press **[ZOOM]**[▷]**[4:SetFactors]** and set the factors.

ex. Find the real zero of $x^3 + x + 1 = 0$
correct to hundredths. (Use both Zoom factors = 4)

Use Zoom In:

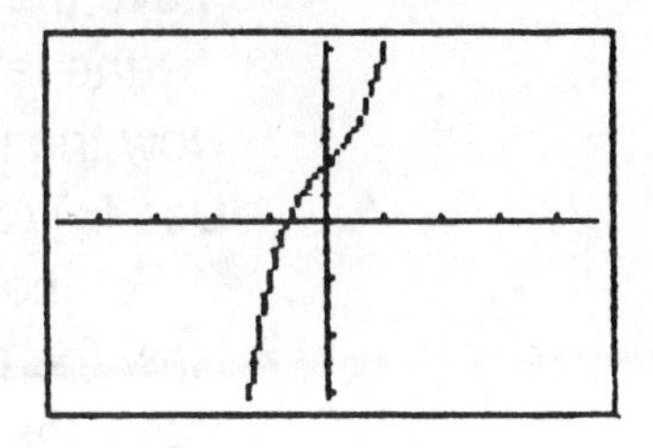

Press **[Y=]** in an empty space
Enter **[x,T,θ,n]**[^] **3** [+]**[x,T,θ,n]**[+] **1**
[ZOOM][4:ZDecimal]
The function crosses the X axis
between -1 and 0.
Press **[WINDOW]**
Set the Xscl = 0.01 **[GRAPH]**
Press **[TRACE]**
Move the cursor near the root.
[ZOOM][2:Zoom In][ENTER] Repeat
[TRACE][ZOOM][2:Zoom In][ENTER] Repeat
[TRACE][ZOOM][2:Zoom In][ENTER]

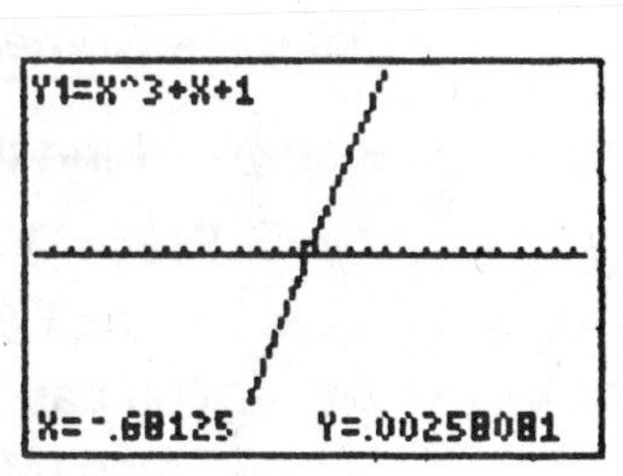

Now the tick marks are visible on the X-axis.

[TRACE]
For $x \approx -.6828125, y \approx -.0011622$
For $x \approx -.68125, y \approx .00258081.$
The value of the zero is $x \approx -.68$ correct to hundredths.

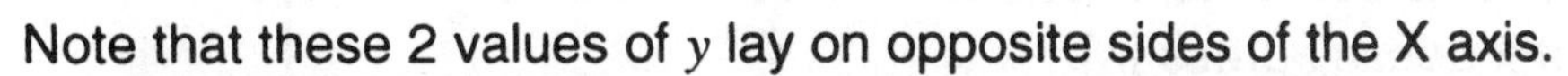
Note that these 2 values of y lay on opposite sides of the X axis.
(These figures depend on the value of the Zoom Factors, these are set at 4).

Use Zoom Box:

Press **[Y=]**, in an empty space enter **[x,T,θ,n]**[^] **3** [+]
[x,T,θ,n][+] **1** **[ZOOM][4:ZDecimal]**
You can see that the function crosses the X axis
between -1 and 0.
Press **[WINDOW]** and set the Xscl = 0.01 **[GRAPH]**
Move the free cursor with the arrow keys until it is above and to
the left of the zero. Press **[ZOOM][1:ZBox][ENTER][ENTER]**
Note that the cursor is now a blinking box.
Use [▷], then [▽] to draw a box with the zero inside.**[ENTER]**
Repeat until the tick marks on the X axis are clearly visible.
Trace to approximate the value $x \approx -.68$

Using the Graph Solve feature:

ex. Find the real zero of $x^3 + x + 1 = 0$ correct to hundredths.

Press **[Y=]** in an empty space enter **[x,T,θ,n]**[^] **3** [+]
[x,T,θ,n][+] **1** **[ZOOM][4:ZDecimal]**
You can see that the function crosses the X axis between -1 and 0.
Press **[2nd][CALC][2:zero]**

The graph is drawn with the request *Left Bound*?
Move the blinking cursor, using [◁], until it is to
the left of the zero, **[ENTER]**
Now the request is *Right Bound*?
Use [▷] to move the cursor to the right
of the zero, **[ENTER]**
Now the request is *Guess*?
Move the cursor as near as possible to the
intersection. Press **[ENTER]**

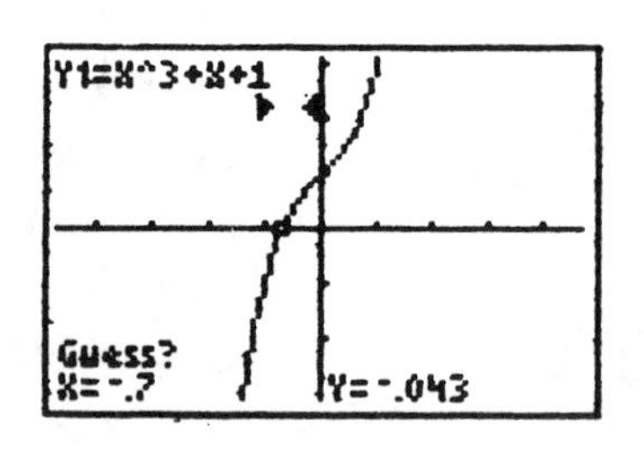

The approximation is $x \approx -.682$, correct to 3 places.

Using the Numeric Solver:

ex. Find the real zero of $x^3 + x + 1 = 0$ correct to hundredths.

Press **[Y=]**, in an empty space enter **[x,T,θ,n]**[^] **3** [+]
[x,T,θ,n][+] **1** **[ZOOM][4:ZDecimal]**
You can see that the function crosses the X axis between -1 and 0.
Press **[MATH]**[△]**[0:Solver]**
The screen should read EQUATION SOLVER.
If it doesn't press [△].
If another equation is entered, press **[CLEAR]**
After **eqn**: **0** = type the equation.
eqn: **0** = $x^3 + x + 1$ is displayed. **[ENTER]**

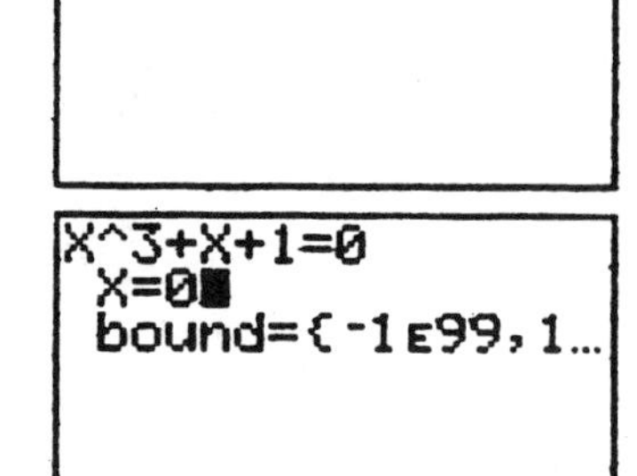

This next screen has the equation on the
top row, and a guess for x on the
second line with a blinking box.
The zero is between -1 and 0, so edit the guess.
Type **0** for a guess, then press **[ALPHA][SOLVE]**
The x value is now $x \approx -.682...$ correct to hundredths.

A guess of -1 or -.5 would have worked as well. Try it.

The blinking square **MUST** be on the x = row
for the solver to compute the answer.

The statement *left – rt* = 0 means the answer is
accurate to the same precision as the calculator.
The answer is an approximation.

Finding the Intersection of 2 graphs:

ex. Find the intersections of $y = x^2 + 2x - 3$ and $y = \frac{x}{2}$.

Press **[Y=]** in empty spaces enter
$Y1$ = **[x,T,θ,n][x^2][+] 2 [x,T,θ,n][–] 3**
$Y2$ = **[x,T,θ,n][÷] 2 [ZOOM][4:Decimal]**
You can see the intersections are in the first and third quadrants. **[TRACE]** to the intersection in Quad. 1. Using [△] and[▽] arrows you can see the coordinates are not "nice" numbers, so use an INTERSECTION SOLVER.
Press **[2nd][CALC][5:intersect]**
the request is *First curve*? the equation of the parabola is marked. **[ENTER]**
the request is *Second curve*? the equation of the line is marked. **[ENTER]**
the request is *Guess*? Move the blinking cursor near the intersection in Quad 1, **[ENTER]**

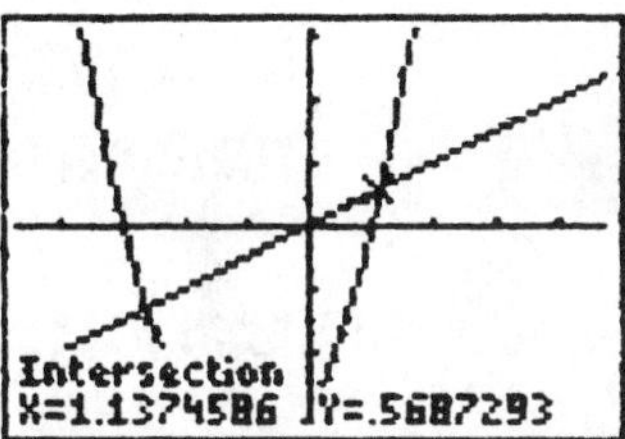

An approximation to the coordinates is
$x \approx 1.137...$ $y \approx .569...$

Using the same procedure in Quad 3,
$x \approx -2.367...$ $y \approx -1.319$

13. Finding Maximum and Minimum values of a Function:

The X value is the location of the extrema.
The Y value is the value of the extrema.
Find the minimum value of $f(x) = x^2 + 3x - 1$

Using Graph Solve:

Press **[Y=]** in an empty space enter
[x,T,θ,n][x^2][+] 3 [x,T,θ,n][–] 1
[ZOOM][4:ZDecimal]
The minimum is off the screen.
Press **[WINDOW]**
Change Ymin to –5.1 and Ymax to 1.1
[GRAPH]
[2nd][CALC][3:minimum]
Move the cursor to the left of the minimum,
Press **[ENTER]**
Move the cursor to the right of the minimum, **[ENTER]**
Then move the cursor to the minimum, **[ENTER]**

The minimum is $\underline{y = -3.25}$

Using the Solver:

This algorithm is iterative and requires a lower and upper x bound for the minimum. These can be estimated from the graph.

For $f(x) = x^2 + 3x - 1$, we can use a lower bound of $x = -2$, and an upper bound of $x = -1$.

Press **[Math][6:fMin(**] In the parenthesis put (expression, variable, lower bound, upper bound)

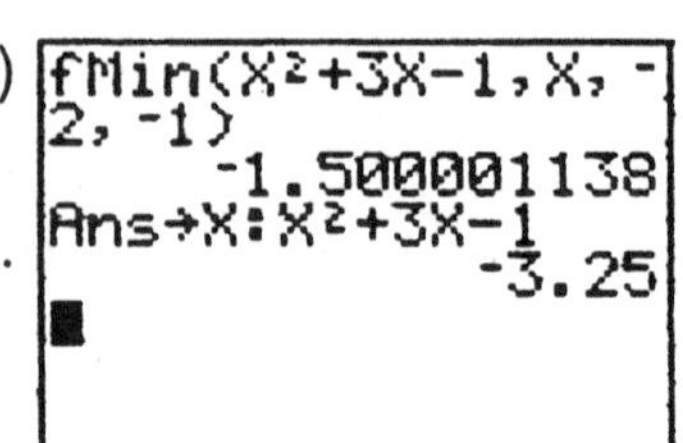

The screen shows **fMin**($x^2 + 3x - 1, x, -2, -1$) **[ENTER]**

The answer is approximately $x \approx$ <u>-1.5000001</u> or $x = -1.5$.

To find the minimum value evaluate $f(-1.5)$.

[(−)]1.5 [STO▸][x,T,θ,n][ALPHA][:]
[x,T,θ,n][x^2][+] **3 [x,T,θ,n][−] 1 [ENTER]**

Read the value $\underline{y = -3.25}$.

14. Statistic Plots:

Entering Data:

Press **[STAT] 1** to display the list editor.
Clear all data from each list by pressing [△]**[CLEAR][ENTER]**
Use the following data:

x	1	2	3	4	5	6	7	8
y	1.1	2.6	3.8	5.1	5.9	7.2	8.2	9

Enter the independent variables, x, in list $L1$.
Highlight the first position in $L1$.
Press **1 [ENTER] 2 [ENTER]** ...
Enter the dependent variables, y, in list $L2$.
Highlight the first position in $L2$.
Press **1.1 [ENTER] 2.6 [ENTER]** ...

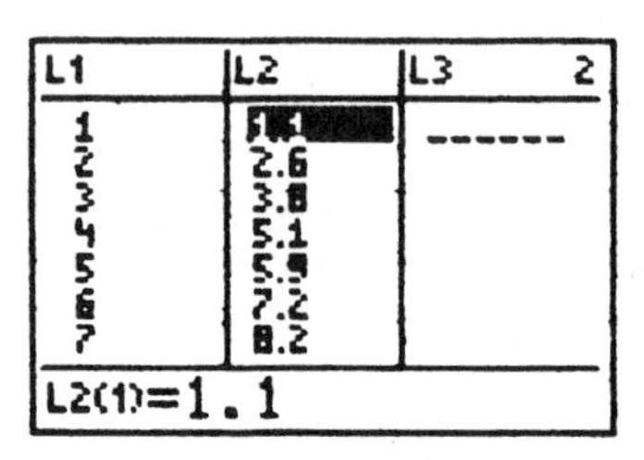

Constructing a Graph:

Press **[Y =**] and turn off or clear any functions.
Press **[2nd][STAT PLOT] 1** to select Plot 1
Press **[ENTER]** to turn ON Plot 1.
Make the following selections, highlight, press **[ENTER]**

Type: Scatter Plot (the first icon)
Xlist: $L1$
Ylist: $L2$

Mark:
Press **[ZOOM][9:ZoomStat]** to see the Scatter Plot

Regression Analysis:

Press **[STAT]**[▹] to display the CALC menu
Select the regression type [**4**:**LinReg**(**ax**+**b**)][**ENTER**]
The regression coefficients and equation are displayed.

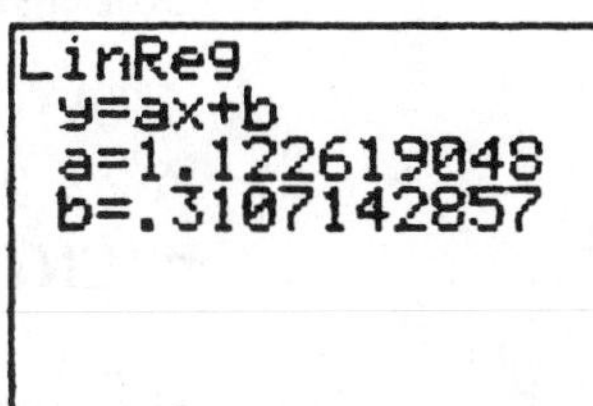

Regression Plot:

A regression analysis is needed before a Plot can be drawn.
Press [**Y**=] to display the equation editor.
Move to an empty space.
Press [**VARS**][**5**:**Statistics**]
[▹][▹][**1**:**RegEQ**][**ENTER**]
The current regression equation is copied to the Y = menu.
Press [**GRAPH**]
Watch the regression line plot through the scatter plot.

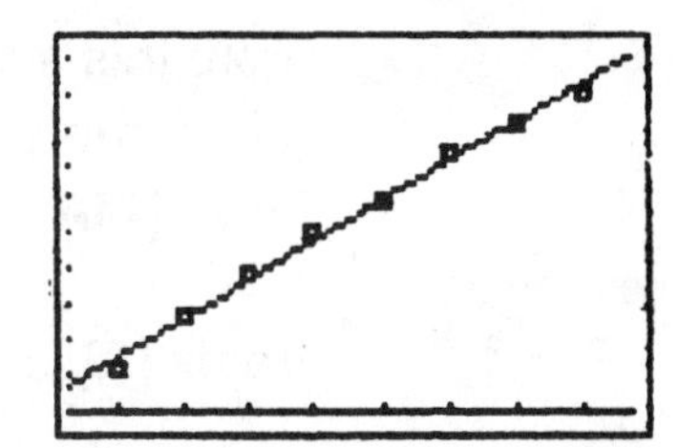

15. Complex Numbers:

Rectangular Complex:

The number $i = \sqrt{-1}$ is **[2nd][i]** located on the bottom row of the keyboard.
To enter 2 + 3i, press **2** [+] **3 [2nd][i]**
To multiply or divide complex numbers, each number must be entered in a parenthesis.

ex. (**2** + **3i**) + (**3** − **5i**) = $\underline{5-2i}$

Polar Complex:

The number a + bi is written $re^{i\theta}$

$r = \sqrt{a^2 + b^2} \quad \theta = \arctan(\frac{y}{x})$

Set Rectangular complex or Polar complex in the MODE menu.

Complex Functions:

Press **[MATH]**[▹][▹] to display the list of functions and operations for complex numbers.

16. Matrices:

Matrices are <u>*stored*</u> by name, edited and used in matrix arithmetic in the MATRX menu on the keyboard.
The menu holds 10 matrices named A - J.

To enter a matrix, press **[MATRX][▹][▹][ENTER]** (EDIT)
Enter the dimensions, rows × columns then fill in the matrix using **[ENTER]** after each entry.

ex. **[MATRX][▹][▹][ENTER] 2 [ENTER] 2 [ENTER]**
A 2 × 2 grid is displayed with element a_{11} highlighted.
1 [ENTER] 2 [ENTER] 3 [ENTER] 4 [ENTER]

[2nd][QUIT] the matrix [A] = $\begin{bmatrix} 1 & 2 \\ 3 & 4 \end{bmatrix}$ is saved.

Matrix *operations* are displayed with **[MATRX][▹]** (MATH)

From this screen press **[ALPHA] B**
The home screen displays **rref(**
Press **[MATRX][1:[A]][)][ENTER]**

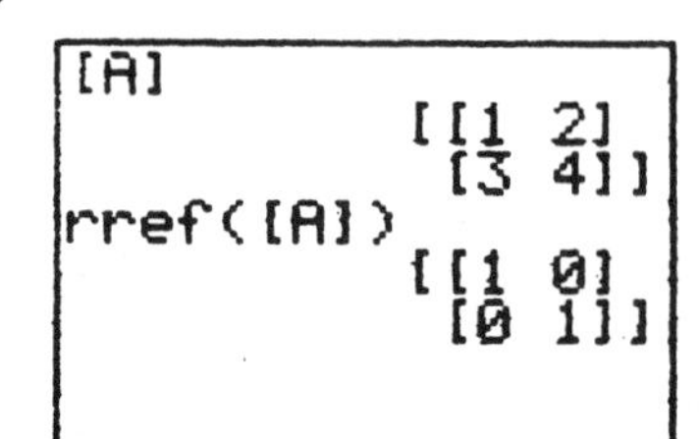

Matrix [A] is row reduced to $\begin{bmatrix} 1 & 0 \\ 0 & 1 \end{bmatrix}$

To save it press **[2nd][ANS][STO▸][MATRX][2:B][ENTER]**
The matrix name [A] or [B] must be pasted onto the home screen from the MATRIX NAMES Menu.

Matrix *arithmetic* +, –, ×,^ is done with the blue operation keys on the keyboard.

Matrix *Row Opreations* are done from the menu **[MATRX]** [▹] MATH submenu,

Press [△]. The row operations are **C**:,**D**:, **E**:, **F**:.
See the TI83 manual for details.

Matrices may be *entered from the keyboard*:

ex. Find the solution to the system of equations:

$$-2x_1 + 3x_2 - 2x_3 = 16$$
$$-5x_1 + 3x_2 - 5x_3 = 22$$
$$x_1 + x_3 = -2$$

<u>*Enter the matrix*</u> $\begin{bmatrix} -2 & 3 & -2 & 16 \\ -5 & 3 & -5 & 22 \\ 1 & 0 & 1 & -2 \end{bmatrix}$

Use [(–)] for the negative numbers
[2nd][[ ][2nd][[ ] -2 , 3 , -2 , 16 [2nd][ ]]
[2nd][[ ] -5 , 3 , -5 , 22 [2nd][ ]]
[2nd][[ ] 1 , 0 , 1 , -2 [2nd][ ]]
[2nd][ ]][ENTER]
[STO▸][MATRX][ENTER][ENTER]

Find the <u>reduced row echelon form</u>:

[MATRX][▷][ALPHA][B][MATRX][1:[A]][ENTER]

<u>*Enter and store*</u>: $\begin{bmatrix} 1 & 1 & 2 & 1 \\ -1 & 1 & 0 & 1 \\ 2 & 1 & 1 & 0 \\ 1 & 3 & 1 & 0 \end{bmatrix}$

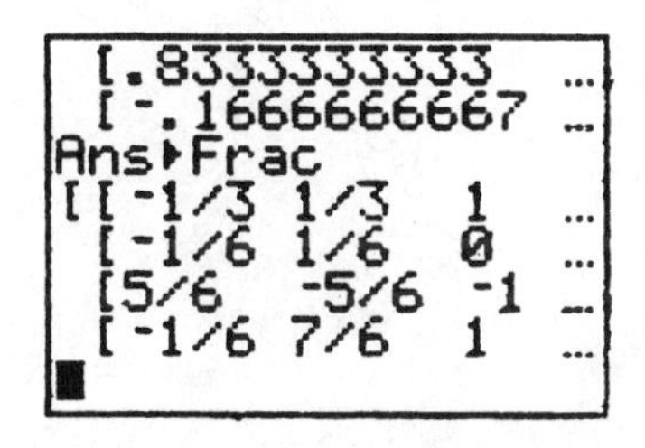

Use [(–)] for a negative number.
[2nd][[ ][2nd][[ ]1 , 1 , 2 , 1 [2nd][ ]]
[2nd][[ ] -1 , 1 , 0 , 1[2nd][ ]]
[2nd][[ ] 2 , 1 , 1 , 0 [2nd][ ]]
[2nd][[ ] 1 , 3 , 1 , 0 [2nd][ ]]
[2nd][ ]][ENTER]
[STO▸][MATRX][1:[A]][ENTER]

Find the <u>Inverse</u>:

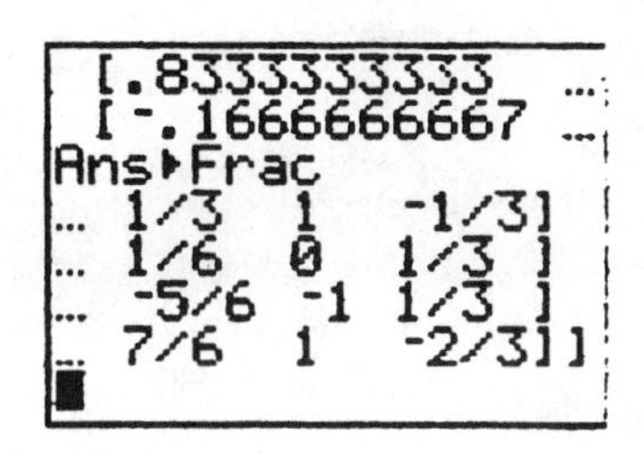

[MATRX][1:[A]][x⁻¹][ENTER]
[STO▸][MATRX][2:[B]]
[ENTER][MATH][1:▸Frac][ENTER]
Use [▷] to view the rest of the matrix.

To <u>check</u> the inverse:

[MATRX][1:[A]][×][MATRX][2:[B]][ENTER]
The answer is the Identity of order 4.

To find the <u>determinent</u> of matrix A:

[MATRX][▷][1:det(][MATRX][1:[A]][)][ENTER]

The answer is <u>6</u>.

Chapter 4 – Using the TI-86 Graphing Calculator.

1. **Notation**:

Keystrokes, except for numbers and commas are in bold brackets **[MATH]**
For a second function $\sqrt{\ }$, press **[2nd]** [$\sqrt{\ }$] the second functions are
printed in yellow on the left over the keys.
For the letter A, press **[ALPHA][A]** the ALPHA symbols are printed in
blue on the right over the keys.
For the letter a, press **[2nd][alpha][A]** the lower case letter is printed.
For functions in the screen menus, press **[2nd]** then the name of the MENU.
The choices are shown on the bottom of the screen.
If an arrow ▸ is shown at the far right, press **[MORE]** to see the other choices.
Each of these choices contains another menu. Use one of the
[F1] through **[F5]** keys to make the first selection,
then use these keys again for the next menu choice.
The original menus are now accessed with
[2nd][M1] through **[2nd][M6]**
To display $\sqrt[3]{2}$, press **[2nd][MATH][F5:MISC][MORE] 3 [F4:** $\sqrt[x]{\ }$ **] 2 [ENTER]**
To leave a screen menu, press **[EXIT]**
To clear the home screen, press **[CLEAR]**
Calculator results are underlined.

2. **Basics**:

To adjust the screen contrast: Press **[2nd]** then [△] to increase
the contrast or **[2nd]** then [▽] to decrease the contrast.
Mode settings: Press **[2nd][MODE]** select all of the settings at the left side
of the screen menu. To change a setting, use the arrow keys
to highlight the desired setting, press **[ENTER]**
Leave these settings unless directed to change a mode setting.

3. **Arithmetic Computations**:

Order of Operations: The order is Algebraic, operations are performed
from left to right. First exponentiation, then multiplication and division,
then addition and subtraction. Parenthesis must be used to change the
algebraic order. Parenthesis must be used around the numerator and
denominator in Algebraic Fractions.
Minus sign: Use [(–)] for negative, [–] for subtraction.
Fractions: to enter fractions use [÷]

To change a decimal to a fraction use
[2nd][MATH][F5:MISC][MORE][F1:▸Frac][ENTER]

To change a fraction to a decimal use **[2nd][CATLG-VARS]**
from A use the up arrow to **▸Dec** press **[ENTER]**

To calculate: press **[ENTER]**

To display a symbol or command from a menu
on the home screen, press **[ENTER]**

ex. $2 \times 3 + 4 \times 5$ **[ENTER]** 26
$2 \times (3 + 4) \times 5$ **[ENTER]** 70
$3 + 4 \div 2$ **[ENTER]** 5
$(3 + 4) \div 2$ **[ENTER]** 3.5
[2nd][MATH][F5:MISC][MORE][F1:▸Frac][ENTER] 7/2

4. Scientific Notation:

From the keyboard use number between 1 and 10 **[EE]** exponent.
The exponent must be a number between -99 and 99.
To Display answers in Scientific Notation change the MODE menu entry.
[2nd][MODE][▷][ENTER][EXIT]
You may want to change the number of decimal places displayed. To
display 3 places, change to FIX 3 press
[2nd][MODE][▽][▷][▷][▷][▷][ENTER][EXIT]
To return to Normal:
[2nd][MODE] Highlight Normal **[ENTER][▽]** highlight Float **[ENTER][EXIT]**

5. Editing:

Before typing ***[ENTER]***

A blinking box determines the current position on the screen.
The arrow keys move the cursor around the screen.

To change an entry move the blinking cursor to the entry, then
type the new entry. It will replace the old entry.

To delete a symbol, move to the symbol, press **[DEL]**

To insert a symbol, move to the symbol after the insertion point,
press **[2nd][INS]** then type the new text.

To erase the input line, press **[CLEAR]**

After typing ***[ENTER]***

Press **[2nd][ENTRY]** the last command is recalled, edit it
as explained above in *Before typing ENTER*

Continue to press **[2nd][ENTRY]** to go back to other commands.

To clear the home screen continue pressing **[CLEAR]** until the screen is cleared.
The home screen can be cleared without clearing any Screen menus.

To exit a MENU, press **[EXIT]** press **[2nd][QUIT]** or enter another MENU.

To delete data in MEMORY, press **[2nd][MEM][F2:Delete]** then delete
from any of the given MENUS, press **[F4:LIST]**
move the selection cursor ▸ to the item, press **[ENTER]**
The contents of the item are gone forever.
Press **[2nd][QUIT]** when done.

To erase a function from the y = screen, highlight any symbol in the function,
press **[CLEAR]**

To erase an entry from the CUSTOM menu, press **[2nd][CATALOG][F4:BLANK]**
then press the menu key corresponding to the entry to delete. It is deleted.

6. **Entering Algebraic Functions**:

Into the y = screen. Press **[GRAPH][F1:y(x)=]** Use the down arrow key
to move to an empty position, or delete a current entry.
Enter the function. using either **[F1:x]** or **[x-VAR]** for x.
It is now available to use in a graph or in a table.
The number of functions displayed is dependent on the amount of free memory.

On the home screen, enter the function. When you press **[ENTER]** it will be
evaluated with the values of the variables that are stored in Memory.

The Catalog is a list of all functions, operations and symbols in the calculator.
Press **[2nd][CATLG-VARS][F1:CATLG]**
You will see a list sorted alphabetically. Press the first letter
of the word you want, use the down arrow to move the
pointer to the entry, press **[ENTER]** and the word is pasted onto
the home screen.

The symbols and some commands are listed after Z. From A, use the up arrow to
display the symbols.

Use **[EXIT]** to leave the CATALOG.

When you enter built in functions, parenthesis must be
included around the argument.

The key **[x-VAR]** prints x. For other letters use **[ALPHA]**

For Greek letters press **[2nd][CHAR][F2:GREEK]**

Special Functions:

When they are entered, they are evaluated with the stored x value.
Absolute value of x: press **[2nd][MATH][F1:NUM][F5:abs][x-VAR]**

Powers, x^5: press **[x-VAR][^] 5**

2^x: press **2 [^][x-VAR]**

e^x: press **[2nd][e^x][x-VAR]** the ^is forced.

Roots, $\sqrt[5]{x}$: press **[2nd][MATH][F5:MISC][MORE] 5 [F4: $\sqrt[x]{\ }$][x-VAR]**

or **[x-VAR][^][(]1 [÷] 5 [)]**

Natural log, $ln(x+5)$: press **[ln][(][x-VAR][+] 5 [)]**

Conjugate Pairs, $\pm\sqrt{4-x^2}$: press **[{] 1 , -1 [}][2nd]**

[$\sqrt{\ }$] 4 [–][x-VAR][x^2][)]

(find the braces { } in the CATALOG or LIST)

Trig function, $tan(3x^2)$: press **[tan][(] 3 [x-VAR][x^2][)][ENTER]**

Inverse trig function, $arcsine(x)$: press **[2nd][$\sin^{-1}$][(][x-VAR][)][ENTER]**

Factorial, 5!: press **5 [2nd][MATH][F2:PROB][F1:!][ENTER]**

Combination, ${}_5C_2$: press **5 [2nd][MATH][F2:PROB][F3: ${}_5C_2$][ENTER]**

The CUSTOM menu provides easy access to functins and commands used frequently. It holds 15 items.

Press **[2nd][CATLG-VARS][F1:CATLG][F3:CUSTM]**

Use the [△] and [▽] or **[M1:PAGE ↓]** or **[M2:PAGE ↑]** to highlight the entry to paste into the CUSTOM menu.

Press the F key where the entry is to be pasted and saved.

That command or function can now be accessed by pressing **[CUSTOM]** on the keyboard using less keystrokes.

ex. Paste the command ▸FRAC into the CUSTOM menu.

[2nd][CATLG-VARS][F1:CATLG][F3:CUSTM] A

Press [△]10 times. The ▸ FRAC command is marked.

Press **[F1]** The command is pasted into the screen menu.

7. Evaluating a function: (checking your answer)

User created Variable names can contain 1-8 characters.

The first character must be a letter, (includes Greek letters), and can not be a built-in variable.

Variable names are case sensitive. AB ≠ Ab ≠ aB ≠ ab.

Numbers or expressions are stored using **[STO▸]** an arrow → is shown on the home screen.

When an algebraic expression is entered, the variables are replaced with the stored constants and a number is displayed.

To evaluate a function on the home screen use the key **[2nd]** [:]to connect commands.

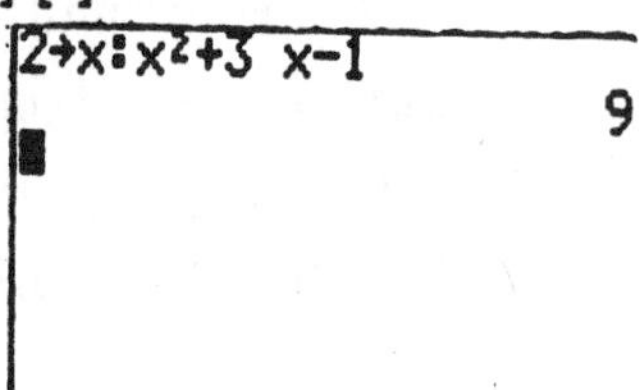

ex. To evaluate x^2+3x-1 for $x=2$,

Press **2 [STO▸][x-VAR][2nd][:][ALPHA]**

[x-VAR][x^2][+] 3 [x-VAR][–] 1 [ENTER]

the result is 9.

If the function is stored in the Y = menu say as y1, it may be evaluated by pressing **[2nd][ALPHA][Y] 1 [ENTER]**

ex. Enter the function $x^2 + 3x - 1$ in the Y = menu as Y1.
Press **[GRAPH][F1:y(x)=]** then
[x-VAR][x²][+] 3 [x-VAR][–] 1 [ENTER][2nd][QUIT]
Store the number 2 in memory x, **2 [STO▸][x-VAR][ENTER]**
Press **[2nd][ALPHA][Y] 1 [ENTER]**
the result is 9.

All expressions containing an x will be evaluated with $x = 2$ until this number stored in the x box is replaced with another number.

If the function is selected in the y= menu, it may be evaluated by pressing **[2nd][MATH][F5:MISC][MORE][EVAL] 2**

If more than one function is selected, the calculator returns a list of the values of each selected function (in order) at the given value of the variable.

8. Angles:

The type of measure for angles is set in the MODE menu.

Set **[2nd][MODE]** to either **Radian** or **Degree** by highlighting the entry and pressing **[ENTER][EXIT]**

After the MODE is set, use the operations in **[2nd][MATH] [F3:ANGLE]** to change the units.

To enter an angle in Degrees while in **Radian** MODE:

ex. Press **[SIN] 30 [2nd][MATH]**
[F3:ANGLE][F1:°][)][ENTER] .5

To enter an angle in Radians while in **Degree** MODE:

ex. Press **[SIN]** $\pi \div$ **6 [2nd][MATH]**
[F3:ANGLE][F2:ʳ][)][ENTER] .5

To change Rectangular Coordinates to Polar (Degree Mode):

ex. Press **[2nd][CPLX][MORE][(]1,1 [)]**
[F1:▸Pr(] [ENTER]
1.4142... is the radius in polar form.
∠45... is the angle in **Degrees**.

If the MODE is Radians, enter the angle in Radians.

9. Building Tables:

The TI 86 has a built in TABLE function used to evaluate functions for different values of x. The algebraic expression to be evaluated is stored in the

Y = menu. Press **[TABLE][F2:TBLST]**, set the initial value for x, and the step size, ΔTbl, between entries, or set ASK to enter a random set of numbers for x.

ex. Enter $x^2 + 3x - 1$ in Y1 in the Y = Menu.

Press **[TABLE][F2:TBLST]**, set the following values:

[(–)] 4 [ENTER] for TblStart

1 [ENTER] for ΔTbl

Highlight **Auto** for Indpnt **[ENTER]**

Press **[F1:TABLE]** to view the table.

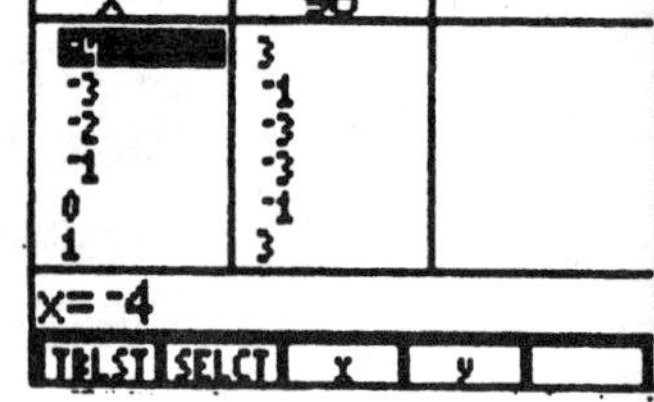

Use the up and down arrows to see more values in the table.

Tables may also be constructed in the STAT menu to examine and evaluate statistical data .

10. Building Lists:

Numbers can be stored in Lists either on the home screen or in the LIST Editor menu.

Lists may be named (stored) using 1 - 8 characters, starting with a letter.

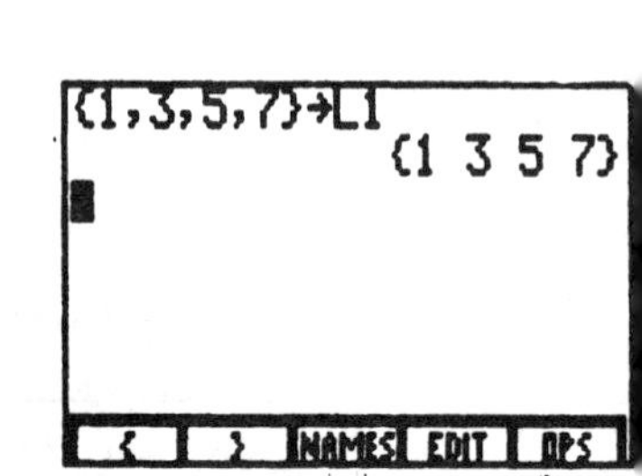

ex. Enter the numbers 1, 3, 5, 7 in list **L1**

[2nd][LIST][F1:{] 1 , 3 , 5 , 7 [F2:}]

[STO▸] [L] [ALPHA] 1 [ENTER]

To display L1 on the home screen, press **[ALPHA][L] 1 [ENTER]**

Arithmetic operations are done using the +,–,×,÷,^ keys, when the lists are the same length.

Other operations are done from the **[2nd][LIST][F5:OPS]** menu.

Special Lists:

[F3:seq(] expression, variable, begin, end increment)

Generate a sequence of numbers using a rule for expression.

ex. **[2nd][LIST][F5:OPS][MORE]**

[F3:seq(][x-VAR][x²] , [x-VAR] , 1 , 5 , 2][)][ENTER]

square the numbers starting with 1, adding 2, ending at 5.

The result is { 1, 9, 25}

ex. **[2nd][LIST][F5:OPS][MORE][MORE]**

[F3:cSum(] (list) returns a list of cumulative sums starting with the first entry. For L1, { 1 , 4 , 9 , 16 }

ex. **[2nd][LIST][F5:OPS][MORE]**

[F1:sum(] (list) returns the sum of all the numbers in the list. For L1, the sum is 16.

ex. **[2nd][LIST][F5:OPS][MORE][MORE]**
[**F4**:Deltal(] (list) returns a list of the differences of consecutive terms. For L1, the list is { 2 , 2 , 2 }
This list is 1 shorter than the original list.

11. Graphing Functions:

The function to be graphed must be entered as one of the $y(x)$= functions in the GRAPH menu. Press **[GRAPH][F1**:**y(x)=]**
Enter **[x-VAR]**[x^2][+] **3** **[x-VAR]**[–] **1** in Y1.

The function must be turned on to graph, therefore, the = must be highlighted.
To turn a function on or off, have the cursor on the function, press [**F5**:**SELCT**].
Each SELCT reverses the current state.

To graph the function press **[2nd][M5**:**GRAPH]**
The function is graphed on the current window settings.

The ZOOM menu has choices to zoom in or out and for built in window sizes.

Press [**F3**:**ZOOM**], press **[MORE][F4**:**ZDECM]**
The window is set to a Friendly window
Each x pixel is 0.1 units.
Unit size on X axis ≠ unit size on Y axis.
Any multiple or translation of these settings also makes a Friendly window.

Press [**F3**:**ZOOM][F4**:**ZSTD**] This is the Standard window, it sets X to [-10,10] and Y to [-10,10]. The units on the X and Y axis are not the same size, so graphs are not their true shape.

Press [**F3**:**ZOOM][MORE][F2**:**ZSQR**] Now the graph is its true shape again. The window settings for the Y axis are left the same, and the X settings have been recalculated so that the units on the X and Y axis are the same. This is called a Square window. It often is not Friendly.

Use ZSQR to get a true shape from a decimal window.

Press [**F3**:**ZOOM][MORE][F1**:**ZFIT**] This is the AutoScale function. After the Xmax, Xmin are set, it finds a Y-range to show a Complete Graph. The Y settings often need adjustment to get a clear picture. With Zoom Fit a Friendly Window stays Friendly.

Press **[2nd][M2**:**WIND]** It displays the current settings for the X and Y axis. These values can be changed by using the arrow keys to highlight the entry, pressing **[CLEAR]** then typing the new value.

Use **[ENTER]** after each change.
(Remember to use (–) for a negative number).
The current values can be multiplied by a constant, or a constant can be added to an entry and the calculator will do the arithmetic.
Press **[F6: GRAPH]** to return to the graph.

Press **[F4: TRACE]** A blinking star cursor appears on the Y axis.
The left and right arrow keys move this cursor along the function graph.
The coordinates that are displayed are points that satisfy the function equation. The Y values are computed from the X value of the pixel.
When more than one function is plotted, the up and down arrows move the cursor vertically between the different plots. These coordinates are only approximations unless the graph is on a Friendly window and the coordinate is a rational number.

To graph a split function $f(x) = \left\{ \begin{array}{ll} x+7, & x \leq -5 \\ 4-x, & x > -5 \end{array} \right\}$

Press **[GRAPH][Y=]** Then enter the 2 functions in empty spaces.
$(x+7)(x \leq -5)$ and $(4-x)(x > -5)$
Enter [(][**x-VAR**][+] **7** [)][÷][(][**x-VAR**]
[2nd][TEST][F4:≤][(–)] **5** [)]in y1.
Enter [(] **4** [–][**x-VAR**][)][÷][(][**x-VAR**]
[2nd][TEST][F3:>][(–)] **5** [)]in y2.
Graph using **[ZOOM][F4: ZSTD]**

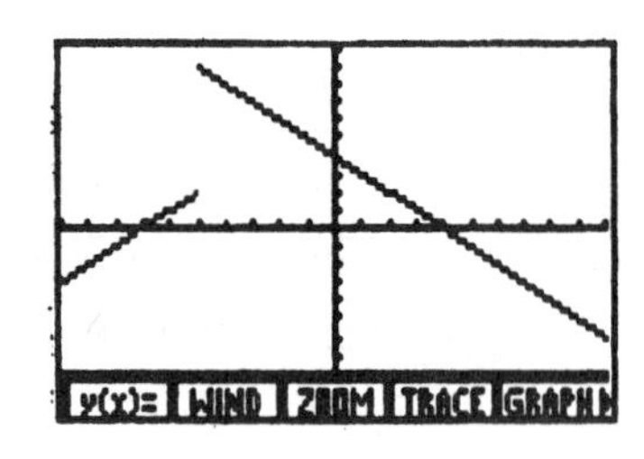

To graph *Parametric Equations*:

Press **[2nd][MODE]** arrow down to **Func**,
Use [▷][▷] to highlight **Param** then press **[ENTER]**
Press **[EXIT][GRAPH][F1: E(t)=]** the menu now reads **xt1**= and **yt1**=
You may want to change the window settings. Now x and y are both functions of the independent variable t.
The settings for $t$min and $t$max determine how much of the graph is plotted.
$t$step determines how many points are plotted.
Enter the functions and press **[GRAPH]**

ex. Write the function $y = x^2 + 3x - 1$ in parametric form.
Let $x = t, \quad y = t^2 + 3t - 1$.
Press **[GRAPH][F1: E(t)=]** Use **[F1: t]** to
Enter **xt1** $= t$
yt1 $= t^2 + 3t - 1$
Press **[2nd][M2: WIND]**
Set **tmin** = [(-)] **5**, **tmax** = **5**, **tstep** = **.5**

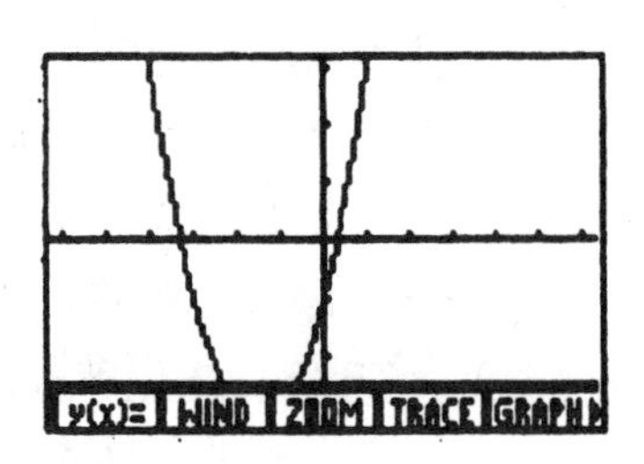

Press **[F3:ZOOM][MORE][F4:ZDECM]** The graph is displayed.
Use **[F4:TRACE]** This is a Friendly Window

To graph *implicit functions* easily,

write the equation in Parametric form.

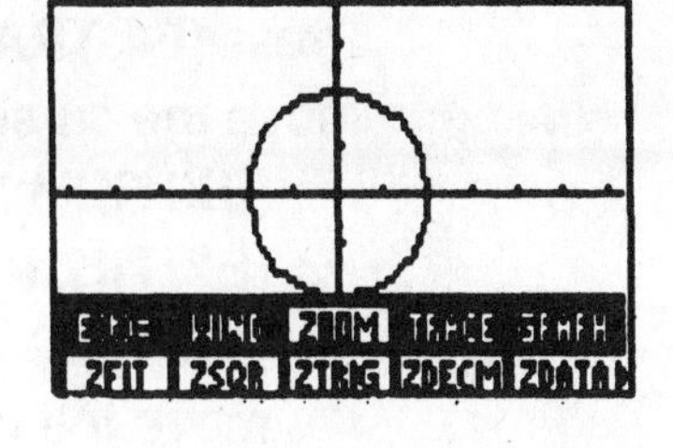

ex. Write the equation $x^2 + y^2 = 4$
in parametric form: $x = 2\cos(t)$, $y = 2\sin(t)$
Press **[GRAPH][E(t)=]**
Enter **xt1** $= 2\cos(t)$, **yt1** $= 2\sin(t)$ **[EXIT]**
[F3:ZOOM][MORE][F4:ZDECM]
[F3:ZOOM][MORE][F2:ZSQR]
The graph is a circle.

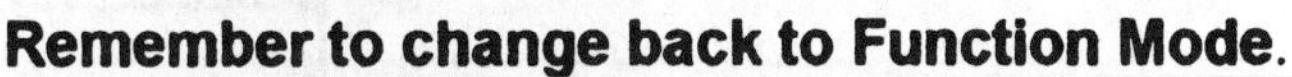

Remember to change back to Function Mode.

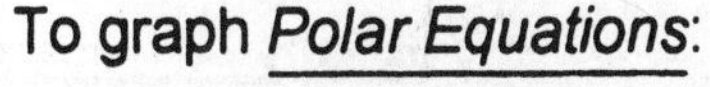

To graph *Polar Equations*:

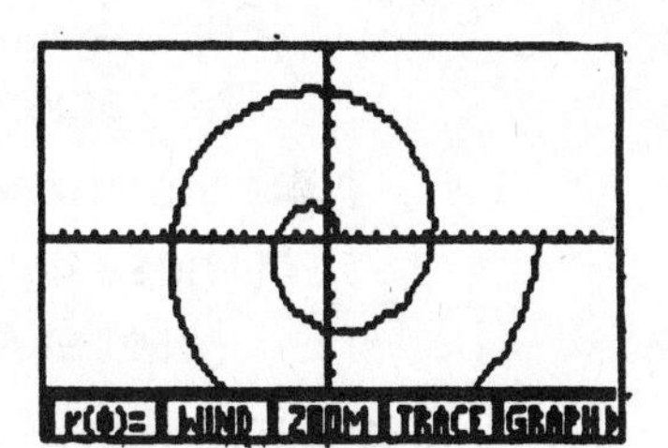

Press **[2nd][MODE]**, arrow down to **FUNC**
Use [▷] to highlight **Pol**, then press **[ENTER]**
[EXIT][GRAPH][F1:$r(\theta)$=]
Enter **r1**=θ, where r = f(θ). Use **[F1:θ]**
Polar is a special case of Parametric,
[2nd][M2:WIND]
Set θ $[0, 4\pi](\frac{\pi}{24})$ set the window [**-10,10**] by [**-10,10**]
[F3:ZOOM][F4:STD]
This is not a friendly window.
[F3:ZOOM][MORE][F2:SQR]
to see the true shape of the graph.

Remember to change back to Function Mode.

12. Solving Equations:

Write the equation in the form $f(x) = 0$
Using Trace and Zoom will give you an approximate answer.
Use either a Zoom box or use Zoom In.

To Set the ZOOM Factors:

Press **[GRAPH][F3:ZOOM][MORE][MORE]**
[F2:ZFACT] and set the factors to xFact=4, yFact=4

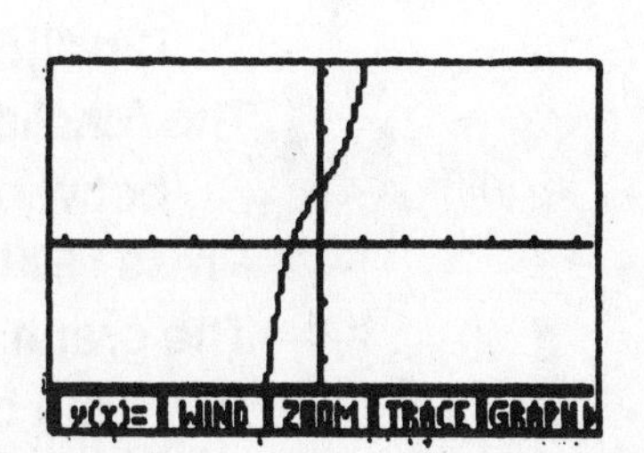

ex. Find the real zero of $x^3 + x + 1 = 0$
correct to hundredths. (Use both Zoom factors = 4)

Use Zoom In:

Press **[GRAPH][y(x)=]**, in an empty space
Enter **[x-VAR][^] 3 [+][x-VAR][+] 1**.
[2nd][M3:ZOOM][MORE][F4:ZDECM]

The function crosses the X axis
between -1 and 0.
Press **[F2:WIND]**
Set the Xscl = 0.01. **[ENTER][F5:GRAPH]**
Press **[F4:TRACE]**,
Move the cursor near the root.
[EXIT][F3:ZOOM][F2:ZIN][ENTER] Repeat
[EXIT][2nd][M4:TRACE]move the cursor near the root,
[EXIT][F2:ZIN][ENTER] Repeat. **[EXIT][2nd][M4:TRACE]**
move the cursor near the root,
[EXIT][F2:ZIN][ENTER]
Now the tick marks are visible on the X-axis.
[F4:TRACE]
For $x \approx -.6828125,\ y \approx -.00116159$
For $x \approx -.68125,\ y \approx .00258081055$.
The value of the zero is $\underline{x \approx -.68}$ correct to hundredths.
Note that these 2 values of y lay on opposite sides of the X axis.
(These figures depend on the value of the Zoom Factors, these are set at 4).

Use Zoom Box:
Press **[GRAPH][y(x)=]**, in an empty space
Enter **[x-VAR]**[^] **3** [+]**[x-VAR]**[+] **1**.
[2nd][M3:ZOOM][MORE][F4:ZDECM]
You can see that the function crosses the X axis between -1 and 0.
Press **[F2:WIND]** and set the Xscl =0.01 **[ENTER][F5:GRAPH]**
Move the free cursor with the arrow keys until it is above and to the left of the zero. Press **[F3:ZOOM][F1:Box][ENTER]**
Note that the cursor is now a blinking box.
Use [▷], then [▽] to draw a box with the zero inside. **[ENTER]**
Repeat until the tick marks on the X axis are clearly visible.
Trace to approximate the value $x \approx -.68$

Using the Graph Solve feature:
ex. Find the real zero of $x^3 + x + 1 = 0$ correct to hundredths.
Press **[GRAPH][y(x)=]**, in an empty space
Enter **[x-VAR]**[^] **3** [+]**[x-VAR]**[+] **1**.
[2nd][M3:ZOOM][MORE][F4:ZDECM]
The function crosses the X axis
between -1 and 0.
Press **[EXIT][MORE][F1:MATH][F1:Root]**.
The graph is drawn with the request *Left Bound*?
Move the blinking cursor, using [◁], until it is to

the left of the zero, **[ENTER]**.

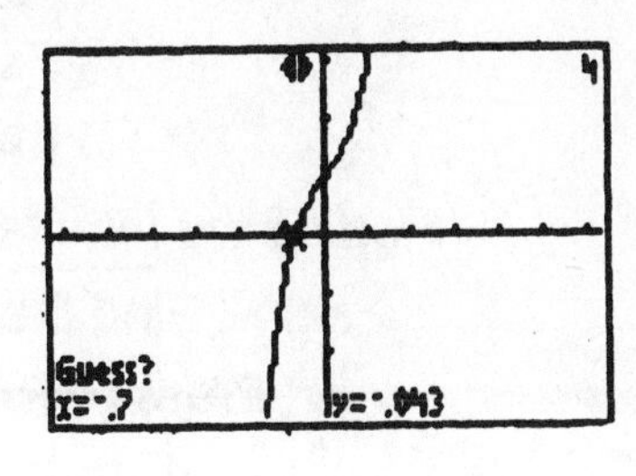

Now the request is *Right Bound*?

Use [▷] to move the cursor to the right of the zero, **[ENTER]**.

Now the request is *Guess*?

Move the cursor as near as possible to the intersection. Press **[ENTER]**.

The approximation is $x \approx -.682$, correct to 3 places.

Using the Numeric Solver.

ex. Find the real zero of $x^3 + x + 1 = 0$ correct to hundredths.

Press **[GRAPH][y(x)=]**, in an empty space

Enter **[x-VAR][^] 3 [+][x-VAR][+] 1**.

[2nd][M3:ZOOM][MORE][F4:ZDECM]

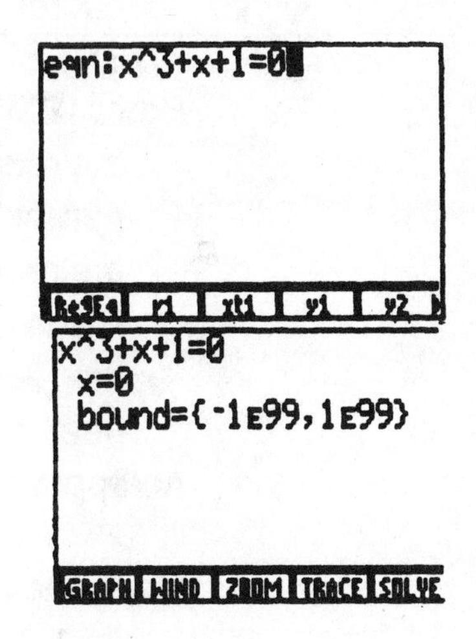

The function crosses the X axis between -1 and 0.

Press **[2nd][SOLVER]**.

The screen should read **eqn**:.

If another equation is entered, press **[CLEAR]**.

After **eqn**: type the equation = 0

eqn: $x^3 + x + 1 = 0$ is displayed.

[ENTER]

This next screen has the equation on the top row, and a guess for x on the second line with a blinking box.

Type **0** for a guess,**[F5:Solve]**

The x value is now $x \approx -.6823278...$

(A guess of -1 would have worked as well. Try it.)

The blinking square **MUST** be on the x = row for the solver to compute the answer.

The statement *left* – *rt* = $1E^{-N}$ means the answer is accurate to the same precision as the calculator.

The answer is an approximation

Using the Polynomial Solver

To find the real and complex zeros of a polynomial with degree between 2 and 20,

use the command POLY.

ex. Find all of the zeros of $x^3 + x + 1 = 0$

Press **[2nd][POLY]** Enter the order **3 [ENTER]**

Fill in the coefficients, **1 [ENTER] 0 [ENTER] 1 [ENTER] 1**

Press **[F5:SOLVE]**

The zeros are displayed as complex numbers.

$x_1 \approx (0.3412, 1.1615)$
$x_2 \approx (0.3412, -1.1615)$
$x_3 \approx (-.6823, 0)$

Finding the Intersection of 2 graphs:

ex. Find the intersections of $y = x^2 + 2x - 3$ and $y = \frac{x}{2}$.

Press **[GRAPH][y(x)=]** in empty spaces enter
$Y1$ = **[x-VAR]**[x^2][+] **2 [x-VAR]**[–] **3**
$Y2$ = **[x-VAR]**[÷] **2 [M3:ZOOM][MORE][F4:ZDECM]**

You can see the intersections are in the first and third quadrants. **[F4:TRACE]** to the intersection in Quad. 1. Using [▵] and[▿] arrows you can see the coordinates are not "nice" numbers, so use an INTERSECTION SOLVER.

Press **[EXIT][MORE][F1:MATH][MORE][F3:ISECT]**
the request is *First curve*? the equation of the parabola is marked. **[ENTER]**
the request is *Second curve*? the equation of the line is marked. **[ENTER]**
the request is *Guess*? Move the blinking cursor near the intersection in Quad 1, **[ENTER]**

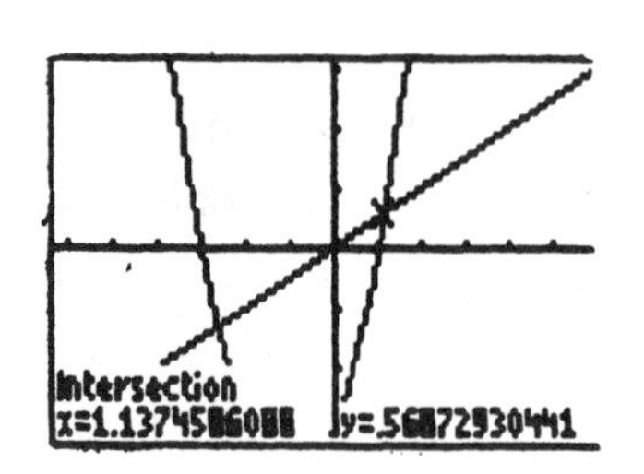

An approximation to the coordinates is
$x \approx 1.137...$ $y \approx .569...$

Using the same procedure in Quad 3,
$x \approx -2.367...$ $y \approx -1.319$

13. Finding Maximum and Minimum values of a Function:

The X value is the location of the extrema.
The Y value is the value of the extrema.
Find the minimum value of $f(x) = x^2 + 3x - 1$

Using Graph Solve:

Press **[GRAPH][y(x)=]** in an empty space
Enter **[x-VAR]**[x^2][+] **3 [x-VAR[–] 1**
[2nd][M3:ZOOM][MORE][F4:ZDECM]
The minimum is off the screen.
Press **[F2:WIND]**
Change Ymin to –5.1 and Ymax to 1.1
[F5:GRAPH][MORE][F1:MATH]
[F4:FMIN].
Move the cursor to the left of the minimum,
Press **[ENTER]**

Then move the cursor to the right, **[ENTER]**
Then move the cursor to the minimum, **[ENTER]**
The minimum is $y = -3.25$ at $x = -1.5$

Using the Solver:

This algorithm is iterative and requires a lower and upper x bound for the minimum. These can be estimated from the graph.

For $f(x) = x^2 + 3x - 1$, we can use a lower bound of $x = -2$, and an upper bound of $x = -1$.

Press **[2nd][CALC][MORE][F1:fMin]** In the parenthesis put (expression, variable, lower bound, upper bound)

The screen shows **fmin**$(x^2 + 3x - 1, x, -2, -1)$ **[ENTER]**
The answer is -1.49999999518... or $x = -1.5$.
To find the minimum value evaluate $f(-1.5)$
The x value is stored in memory.

[x-VAR][x^2][+] **3 [x-VAR]**[–] **1 [ENTER]**
read the value $y = -3.25$.

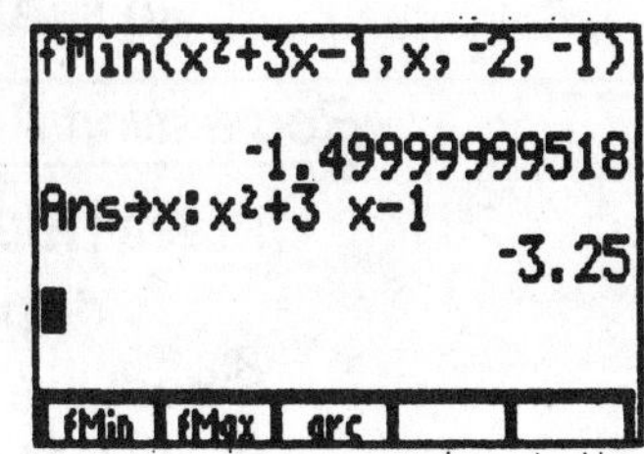

14. Statistic Plots:

Entering Data:

Press **[2nd][STAT][F2:EDIT]** to display the list editor.
Clear all data from each list by pressing [▵]**[CLEAR][ENTER]**
Use the following data:

x	1	2	3	4	5	6	7	8
y	1.1	2.6	3.8	5.1	5.9	7.2	8.2	9.0

Enter the independent variables, x, in list $L1$.
Highlight the first position in $L1$.
Press **1 [ENTER] 2 [ENTER]** ...
Enter the dependent variables, y, in list $L2$.
Highlight the first position in $L2$.
Press **1.1 [ENTER] 2.6 [ENTER]**
[EXIT]

Constructing a Graph:

Press **[GRAPH][Y =]** and turn off **[F5:SELCT]** or clear **[CLEAR]** any functions
Press **[2nd][STAT][F3:PLOT]** to select a Plot
Press **[F1:PLOT1][ENTER]** to turn ON Plot 1.
Make the following selections, use [▿]
If the selections are correct, skip the setup.

Type: Scatter Plot
[F1:SCAT][ENTER]
Xlist Name=xStat
[F1:xStat][ENTER]
Ylist: Name=yStat
[F2:yStat][ENTER]
Mark= ∘,
[F1:∘][ENTER]
[EXIT][EXIT]
to plot on an appropriate screen.
[GRAPH][ZOOM][MORE][F6:ZDATA]
to see the Scatter Plot

Regression Analysis:

Press **[EXIT][2nd][STAT][F1:CALC]]**
to display CALCULATE menu
Select the regression type
[F3:LinR][ENTER],
The regression coefficients and
equation are displayed.
Press **[CLEAR][EXIT][EXIT]**
to return to the home screen.

LinReg
y=a+bx
a=.310714286
b=1.12261905
corr=.996922889

Regression Plot:

A regression analysis is needed before a Plot can be drawn.
Press **[GRAPH][F1:y(x)=]** to display the equation editor.
Move to an empty space.
Press **[2nd][STAT][F5:VARS][MORE][MORE]**
[F2:RegEq][ENTER]
The current regression equation is copied
to the Y = menu as RegEq
Press **[EXIT][2nd][M5:GRAPH]**
Watch the regression line plot through
the scatter plot.

15. Complex Numbers:

Rectangular Complex

The complex number $a + bi$ is entered (a, b)
To enter 2 + 3i, press [(] **2** [,] **3** [)] seen as (2,3)
To multiply or divide complex numbers, each number
must be entered in a parenthesis.

ex. (2, 3)×(3, -5) = (21, -1)

Polar Complex:

The number a + bi is written $re^{i\theta}$

$r = \sqrt{a^2 + b^2}$ $\theta = \arctan(\frac{y}{x})$

The number is entered $(r\angle\theta)$

ex. To enter $3e^{i\pi}$ enter[(] **3 [2nd]**[∠]**[2nd]**[π][)]

To display all results in Polar form,
set Polar complex in the MODE menu.

Complex Functions:

Press **[2nd][CPLX]** to display the list of functions and operations for complex numbers.

16. Matrices:

Matrices are *stored* by name, edited and used in matrix arithmetic in the MATRX menu on the keyboard.

To enter a matrix in the matrix editor,

press **[2nd][MATRX][F2:EDIT]**

Stored matrices are available from the screen menu.

Name a new matrix with 1 to 8 characters,
starting with a letter A, ALPHA MODE is ON
[ENTER] the row dimension is highlighted.

You may overwrite another Matrix.

Enter the dimensions, rows × columns then fill in the matrix using **[ENTER]** after each entry. Use **[ENTER]** to skip an entry.

ex. **[2nd][MATRX][ENTER] 2 [ENTER] 2 [ENTER]**

A 2 × 2 grid is displayed with element a_{11} highlighted.

1 [ENTER] 2 [ENTER] 3 [ENTER] 4 [ENTER]

[2nd][QUIT] the matrix [A] = $\begin{bmatrix} 1 & 2 \\ 3 & 4 \end{bmatrix}$ is saved.

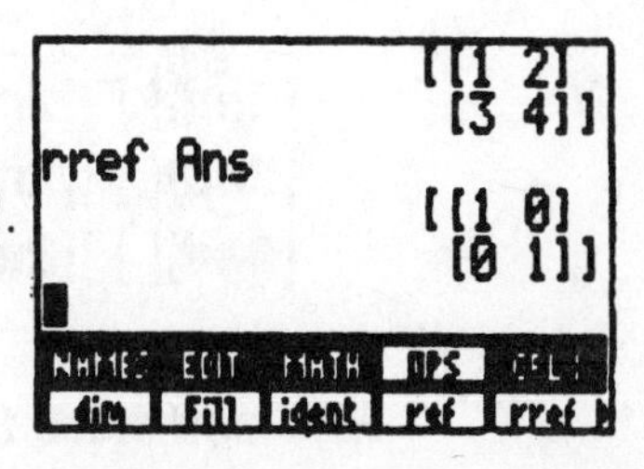

Matrix *operations* are displayed with **[2nd][MATRX][F4:OPS]**.

From this screen press **[F5:rref]**

The home screen displays **rref(**

Press **[EXIT][F1:NAMES]F1:A][ENTER]**

Matrix [A] is row reduced to $\begin{bmatrix} 1 & 0 \\ 0 & 1 \end{bmatrix}$

To save it press**[STO▸][B][ENTER]**

The matrix name [A] or [B] must be pasted onto the home screen from the MATRIX NAMES Menu.

Matrix arithmetic +, –, × is done with the arithmetic keys on the keyboard.

Matrix Row Opreations are done from the menu

[2nd][MATRX][F4:OPS][MORE]

See the TI86 manual for details.

Matrices may be entered from the keyboard.

Example: *Find a solution to the system:*

$$-2x_1 + 3x_2 - 2x_3 = 16$$
$$-5x_1 + 3x_2 - 5x_3 = 22$$
$$x_1 \quad + x_3 = -2$$

Enter the matrix $\begin{bmatrix} -2 & 3 & -2 & 16 \\ -5 & 3 & -5 & 22 \\ 1 & 0 & 1 & -2 \end{bmatrix}$

Use [(–)] for the negative numbers

[2nd][[ ][2nd][[ ] -2 , 3 , -2 , 16 [2nd][ ]][2nd][[ ]
-5 , 3 , -5 , 22 [2nd][ ]][2nd][[ ] 1 , 0 , 1 , -2 [2nd][ ]]
[2nd][ ]][ENTER][STO▸][A][ENTER]

Find the reduced row echelon form:

[2nd][MATRX][F4:OPS][F5:rref][EXIT]
[F1:NAMES][F1:A][ENTER]

Example:

Enter and store the matrix: $\begin{bmatrix} 1 & 1 & 2 & 1 \\ -1 & 1 & 0 & 1 \\ 2 & 1 & 1 & 0 \\ 1 & 3 & 1 & 0 \end{bmatrix}$

[2nd][[ ][2nd][[ ]1 , 1 , 2 , 1 [2nd][ ]][2nd][[ ] -1 , 1 , 0 , 1
[2nd][ ]][2nd][[ ] 2 , 1 , 1 , 0 [2nd][ ]][2nd][[ ] 1 , 3 , 1 , 0
[2nd][ ]][2nd][ ]][ENTER][STO▷][A][ENTER]

Find and store the inverse

[2nd][MATRX][F1:NAMES][F1:A][2nd][x^{-1}][ENTER]
[STO▷][B][ENTER]
[2nd][MATH][F5:MISC][MORE][F1:▸Frac][ENTER]

To check the inverse:

[2nd][MATRX][F1: **NAMES][F1**: A][×][**F2**: B][**ENTER**]

To find the *determinant* *of matrix A*:

[2nd][MATRX][F3: **MATH][F1**: **det]**

[2nd][M1: **NAMES][F1**: A][**ENTER**]

The answer is 6

Chapter 5

Using the TI-89 Graphing Calculator.

1. **Notation**:

Keystrokes, except for numbers and commas are in bold brackets [**MATH**]

For a second function √‾, press [**2nd**] [√‾] the second functions are printed in yellow on the left over the keys.

For the letter A, press [**alpha**][**A**], the Alpha symbols are printed in purple on the right over some keys.

For functions in the screen menus ≤ press [**2nd**] [**MATH**][**8**:**Test**] [**3**:≤] To execute a screen menu function say Test, you may press the number **8** or use [▽] to move down and highlight the entry [**8**:**Test**] and press [**ENTER**]

The green diamond key [◇] is used for functions on the keyboard that are in green on the right over some keys. It also is used with other keys to activate menus and applications.

The shift key [⇑] types an uppercase letter for the letter typed after it. It is also used for highlighting in editing.

Calculator results are underlined.

2. **Basics**:

To adjust the screen contrast:

Press and hold both [◇] and [+] to increase the contrast (darken).

Press and hold both [◇] and [–] to decrease the contrast (lighten).

About the *Home Screen*:

The following illustration of the Home Screen shows the entry line and the history area.

Entry/Answer pairs in the history area are displayed in "pretty print".

Mode settings: Modes control how numbers and graphs are displayed and interpreted.

Press [**MODE**] select the following settings. Leave these settings unless directed to change a mode setting.

Graph	Function
Display Digits	Float 6
Angle	Radian
Exponential Format	Normal
Complex Format	Real
Pretty Print	On
Split Screen	Full
Exact/Approx	Auto
Base	Dec

(Auto gives exact answers if possible, otherwise a decimal)

Press [**ENTER**] to leave the MENU

3. **Arithmetic Computations**:

Order of Operations: The order is Algebraic, operations are performed from left to right. First exponentiation, then multiplication and division, then addition and subtraction.

Parenthesis must be used to change the algebraic order.

Minus sign: Use [(–)] for negative, [–] for subtraction.

Fractions: to enter fractions use [÷]

To change a fraction to a decimal use [◇][**ENTER**]

To change back to a fraction use [**ENTER**]

To calculate: press [**ENTER**]

ex. $2 \times 3 + 4 \times 5$ [**ENTER**] 26

$2 \times (3 + 4) \times 5$ [**ENTER**] 70

$3 + 4 \div 2$ [**ENTER**] 5

$(3 + 4) \div 2$ [**ENTER**] 7/2

The answer is in fractional form because the "Pretty Print" mode is on.

To get a decimal answer press [◇][**ENTER**] 3.5

4. **Scientific Notation**:

From the keyboard type the number between 1 and 10 [**EE**] exponent.
The exponent must be an integer between -99 and 99.
To display answers in Scientific Notation change the MODE menu entry.

[**MODE**] use [▽] to Exponential Format [▷] [**2:SCIENTIFIC**] [**ENTER**][**ENTER**]

You may want to change the number of decimal places displayed. To display 3 places, change to FIX 3 in the MODE menu.
To return to Normal

[**MODE**] use [▽] to Exponential Format [▷] [**1:NORMAL**]

[ENTER][ENTER]

5.. **Editing (on the Entry Line):**

The TI-89 has both an insert and overtype mode. Insert is default.

To change between insert and overtype press **[2nd][INS]**

To move the cursor

To the beginning of an expression press **[2nd]**[◃]

To the end of an expression press **[2nd]**[▹]

To Delete

The symbol to the left of the cursor press [←]

The symbol to the right of the cursor press [◇][←]

All symbols to the right of the cursor press **[CLEAR]**

The entire line if the cursor is at the beginning or the end of the entry line. Press **[CLEAR]**

The entire line if the cursor is in the middle of the entry line. Press **[CLEAR][CLEAR]**

To highlight Multiple Characters

Move the cursor to the left of the symbols.
Hold [↑] and press [▹] for each symbol to be highlighted.
Retype the symbols or press [←] to delete them.
To remove a highlight use [▹] or [◃] to move over the highlighted area.

To highlight the Entry Line, press **[ENTER]**

To edit an entry in the history area, press **[2nd][ENTRY]** repeatedly until it moves back to the entry line.

To clear the history area press **[F1][8:Clear Home]**

To exit a MENU press **[ESC]** or **[2nd][QUIT]**

To erase a function from the Y= screen press **[CLEAR]**

To clear a history pair, highlight the entry, press **[CLEAR]**

To return to the Home Screen, press **[HOME]**

6. **Entering Algebraic Expressions:**

In the Y = Menu

To display the Y = Menu press [◇]**[Y=]**
Enter the function. It is now available to use as a graph or table.
This menu holds 99 items.

On the Home Screen, enter the function. When you press **[ENTER]** it will be evaluated with the values of the variables that are stored in Memory.

The Catalog is a list of all functions, operations and symbols in the calculator.

Press [**CATALOG**] you will see a list sorted alphabetically. Press the first letter of the word you want, use the down arrow to move the pointer to the entry, press [**ENTER**] and the word is pasted onto the home screen.

The symbols are listed after Z, use [△] from A

When you enter built in functions, left parenthesis are included. Complete the input variable, then type the right parenthesis.

The key [**x**] prints x in Function mode, [**y**] prints y, [**z**] prints z, for other letters use the [**alpha**] key.

To change exact answers to decimal [◇][**ENTER**]

Special Functions:

Absolute value of x: press [**2nd**][**MATH**][▷][**2**:**abs**][**x**]

Powers x^5: press [**x**][^] **5**

2^x: press **2** [^][**x**]

e^x: press [◇][**x**][)] the ^and (are forced.

Roots, $\sqrt[5]{x}$: press [**x**][^][(]**1** [÷] **5**[)]

Natural log, $\ln(x+5)$: press [**2nd**][ln][**x**][+] **5**

Trig function, $\tan(3x^2)$: press [**tan**] **3** [**x**][^] **2**

Inverse trig function, arcsine(x): [◇][**sin**$^{-1}$][**x**][)]

Factorial, 5!: press **5** [**MATH**][**7**:**Probability**][**ENTER**][**1**:**!**]

Combinations, ${}_5C_2$: press **5** [**MATH**][**7**:**Probability**][**ENTER**] [**3**:${}_nC_r$] **2**

7. **Evaluating a function**: (**checking your answer**)

Numbers can be stored as variables. A variable name can have 1-8 symbols (letters or numbers), the first one can not be a number.

Numbers are stored in these positions using the command [**STO▸**] an arrow → is shown on the home screen.

Single letter Variables will be used in this text.

When an algebraic expression is entered, the variables are replaced with the stored constants and a number is displayed.

To evaluate a function on the home screen use the key [**2nd**][:]to connect commands.

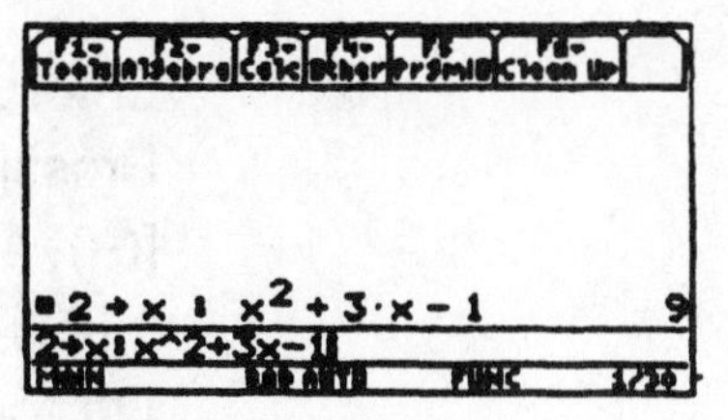

ex. To evaluate $x^2 + 3x - 1$ for $x = 2$,
Press **2 [STO▸][x][2nd]**[:]
[x][x^2][+] **3 [x]**[–] **1 [ENTER]**
the result is 9.

All expressions containing an x will be evaluated with x = 2 until this number in the x box is replaced with another number.

To clear single variables press **[2nd][F6][1:Clr a-z]**

8. Angles:

The type of measure for angles is set in the MODE menu.
Set [**MODE**] to either **Radian** or **Degree** by highlighting Angle and pressing [▷][**1:RADIAN**] or [**2:DEGREE**][**ENTER**]
After the MODE is set, use the operations in [**2nd**][**MATH**] [**2:Angle**] to change the units.

To enter an angle in Degrees while in **Radian** MODE:
ex. Press [**SIN**] **30** [**2nd**][**MATH**][**2:Angle**][**1:°**][)][**ENTER**] .5

To enter an angle in Radians while in **Degree** MODE:
ex. Press [**SIN**] $\pi \div$ **6** [**2nd**][**MATH**][**2:Angle**][**2:ʳ**][)][**ENTER**] .5

To change Rectangular Coordinates to Polar (Degree Mode):
ex. Press [**2nd**][**MATH**][**2:Angle**][**5:R▸Pr(**] **1,1** [)][**ENTER**]
1.4142... is the radius in polar form.
Press [**2nd**][**MATH**][**2:Angle**][**6:R▸Pθ(**] **1,1** [)][**ENTER**]..
45 is the angle in **Degrees**.

To change Polar Coordinates (Degree Mode) to Rectangular:
ex. Press [**2nd**][**MATH**][**2:Angle**][**3:P▸Rx(**] $\sqrt{2}$, **45** [)][**ENTER**] 1
is the x coordinate in rectangular coordinates.
Press [**2nd**][**MATH**][**2:Angle**][**4:P▸Ry(**] $\sqrt{2}$, **45** [)][**ENTER**] 1
is the y coordinate in rectangular coordinates.
If the MODE is Radians, enter the angle in Radians.
[**HOME**] returns to Home Screen.

9. Building Tables:

The TI-89 has a built in table function used to evaluate functions for different values of x. The algebraic expression to be evaluated is stored in the Y = menu.
In the TABLE SET menu, set the initial value for x, and the step sixe between entries, or set ASK to enter a random set of numbers for x.

ex. Enter $x^2 + 3x - 1$ in Y1 in the Y = menu.
Press[◇][**TBLSET**] set the following values
[(–)] **4 [ENTER]** for TblStart
1 [ENTER] for ΔTbl
Off for Graph <-> Table
Auto for Independent
Press [**ESC**] when finished.
Press [◇][**TABLE**] to view the table.

Use the up and down arrows to see more values in the table.
To return to the Home Screen, press [**HOME**]
Tables may also be constructed in the STAT menu to examine and evaluate statistical data .

10. Building Lists:

Numbers can be stored in Lists either on the Home Screen or in the Data/Matrix Editor menu.
Lists may be named (stored) by naming them.
ex. From the Home Screen,enter the numbers 1, 3, 5, 7 in list **L1**
[2nd][{] 1 , 3 , 5 , 7 [2nd][}][STO▸][⇑] L 1 [ENTER]
Lists from the keyboard are stored in the Data/Matrix Editor menu.
To display L1 on the home screen, press [⇑] **L 1 [ENTER]**
Arithmetic operations are done using the +,–,×,÷,^ keys, when the lists are the same length.
Other operations are done from the **[2nd][MATH][3:List]** menu.

Special Lists:

[2nd][MATH][3:List][1:seq(] expression, variable, begin, end increment)
Generate a sequence of numbers using a rule for expression.
ex. **[2nd][MATH][3:List]**
[1:seq(][x][^] 2, [x] , 1 , 5 , 2][)][ENTER]
square the numbers starting with 1, adding 2, ending at 5.
The result is { 1, 9, 25}

[2nd][MATH][3:List][7:cumSum(] (list) returns a list of cumulative sums starting with the first entry.
ex. For L1, { 1 , 4 , 9 , 16 }

[2nd][MATH][3:List][6:sum(] (list) returns the sum of all the numbers in the list.
ex. For L1, the sum is 16.

11. Graphing Functions:

The function to be graphed must be entered as one of the 99 functions
in the Y = menu. Enter $x^2 + 3x - 1$ in Y1.
The function must be turned on to graph. The ✓ must be highlighted.
To turn the function on or off, highlight the function and press [**F4**].
Each time [**F4**] is pressed ✓ reverses the current state.
To graph the function press [◇] [**GRAPH**]
The functions with a ✓ are graphed on the current window settings.
The ZOOM menu [**F2:Zoom**] has choices to zoom in or out and
for built in window sizes.
Press [**F2:Zoom**], press [**4:ZoomDec**] The window is set to
a Friendly window. Each pixel is 0.1 units, and the graph is its
true shape. Unit size on X axis = unit size on Y axis. Any multiple
of these settings also makes a Friendly window.
Press [**ZOOM**][**6:ZoomStd**] This is the Standard window, it sets
X to [-10,10] and Y to [-10,10]. The units on the X and Y axis are not the
same size, so graphs are not their true shape.
Press [**ZOOM**][**5:ZSquare**] Now the graph is its true shape again.
The window settings for the Y axis are left the same, and the X settings
have been recalculated so that the units on the X and Y axis are the
same. This is called a Square window.
Press [◇] [**WINDOW**] It displays the current settings for the X and
Y windows. These values can be changed by using the arrow keys to
highlight the entry, pressing [**CLEAR**] then typing the new value.
Use [**ENTER**] after each change.
(Remember to use (–) for a negative number).
Press [◇] [**GRAPH**] to return to the graph.
Press [**F3:TRACE**] A blinking star cursor appears on the Y axis.
The left and right arrow keys move this cursor along the function graph.
The coordinates that are displayed are points that satisfy the function
equation. The Y values are computed from the X value of the pixel.
When more than one function is plotted, the up and down arrows move
the cursor vertically between the different plots. These coordinates
are only approximations unless the graph is on a Friendly window
and the coordinate is a rational number.

To graph a split function $f(x) = \left\{ \begin{array}{ll} x+7, & x \le -5 \\ -x+4, & x > -5 \end{array} \right\}$

Press [**Y=**] Then enter the function in an empty space.
[**2nd**][**alpha**][**w**][**h**][**e**][**n**][**alpha**][(] $x \le -5, x+7, -x+4)$
The symbol [≤] is in [**CATALOG**]

The vertical line between the segments is a technology error.

To graph Parametric Equations press **[MODE]** from **Function**

use [▷] to highlight **Parametric**, then press **[ENTER]**

Press[◇] **[Y=]** the menu now reads $\mathbf{X}_{1T}$= and $\mathbf{Y}_{1T}$=

You may want to change the window settings. Now X and Y are both functions of the independent variable T.

The settings for Tmin and Tmax determine how much of the graph is plotted.

Tstep determines how many points are plotted.

Enter the functions and press **[GRAPH]**

ex. Write the function $y = x^2 + 3x - 1$ in parametric form.

Let $x = t, y = t^2 + 3t - 1$.

Enter $\mathbf{X}_{1T} = t$ and $\mathbf{Y}_{1T} = t^2 + 3t - 1$

Press **[WINDOW]** Set **Tmin** = (-)5, **Tmax** = **5**.

Press [◇]**[GRAPH]** The graph is displayed.

To graph *implicit functions* easily, write the equation in Parametric form.

ex. Write the equation $x^2 + y^2 = 4$ in parametric form:

$x = 2\cos(t), y = 2\sin(t)$.

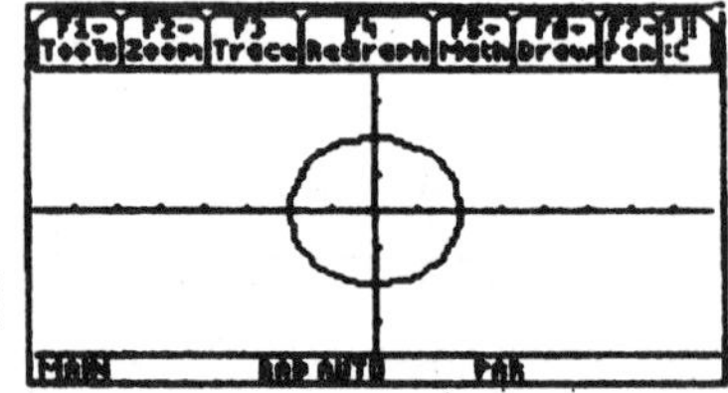

The graph is a circle.

If the circle is not round, press **[F2][4:ZoomDec]**

The regraphed circle is round.

Remember to change back to Function Mode.

To graph *Polar Equations*: press **[MODE]** **Function**

Use [▷]**[3:POLAR]** press **[ENTER]**

Press [◇]**[Y=]** Enter **r1**=θ (θ is on the keyboard.)

[F2:Zoom][4:ZoomDec]

Change θmax. to 6π.

[F2:Zoom][4:ZoomDec]

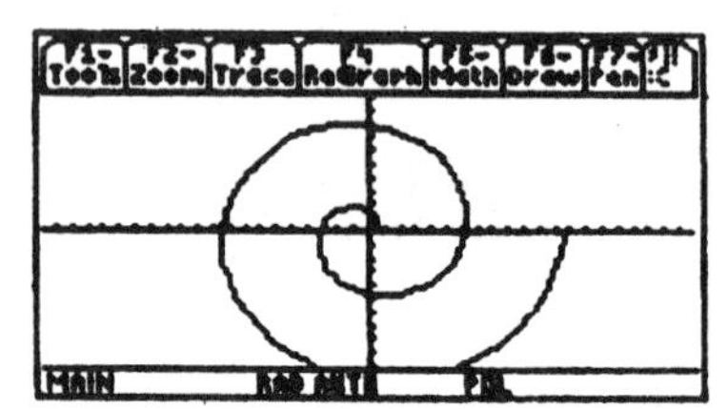

Remember to change back to Function Mode.

12. Solving Equations:

Write the equation in the form $f(x) = 0$

Using Trace and Zoom will give you an approximate answer. Use either a Zoom box or use Zoom In.

ex. Find the real zero of $x^3 + x + 1 = 0$ correct to hundredths.

Use Zoom In:

Press [◇] **[Y=]**

move to Y1, press **[CLEAR]**

Enter [x][^] 3 [+][x][+] 1.
[F2:Zoom][4:ZoomDec]
You can see that the function crosses
the X axis between -1 and 0.
Press **[F3:Trace]** move the cursor near the root.

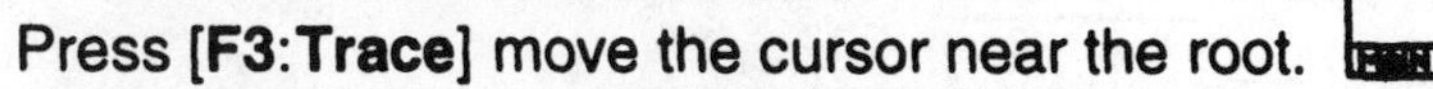

[F2:Zoom][2:Zoom In][ENTER] Repeat
[F3:Trace] move the cursor near the root.
[F2:Zoom][2:Zoom In][ENTER] Repeat
[F3:TraceE] move the cursor near the root.
[F2:Zoom][2:Zoom In][ENTER]
[Trace]
For x = -.682813, y = -.001162
For x = -.679688, y = .006314
The value of the zero is
x ≈ -.68 correct to hundredths.

Note that these 2 values of y lay on opposite sides of the X axis.
You may not get exactly the same numbers.
(These figures depend on the value of the Zoom Factors, these are set at 4).

Use Zoom box:

Press[◇] **[Y=]** in Y1, **[CLEAR]**, then enter **[x][^] 3** [+]
[x][+] 1 [F2:Zoom][4:ZoomDec]
You can see that the function crosses the X axis
between -1 and 0.
Move the free cursor with the arrow keys until it is above and to
the left of the zero. Press **[F2:Zoom][1:ZoomBox][ENTER]**
Note that the cursor is now a blinking box.
Use [▷], then [▽] to draw a box with the zero inside.
Repeat until the tick marks on the X axis are clearly visible.
Trace to approximate the value x ≈ -.68

Using the Graph Solve feature:

ex. Find the real zero of $x^3 + x + 1 = 0$ correct to hundredths.
Press[◇] **[Y=]**, in an empty space enter **[x][^] 3** [+]
[x][+] 1 [F2:Zoom][4:ZoomDec]
You can see that the function crosses the X axis between -1 and 0.
Press **[F5:Math][2:Zero]**
The graph is drawn with the request *Lower Bound?*
Move the blinking cursor with [◁] until it is to the left of the zero,
[ENTER]
Now the request is *Upper Bound?*
Use [▷] to move the cursor to the right of the zero, **[ENTER]**

On the bottom of the screen, read

Zero

xc: -.682328 *yc*:0

The approximation is x = -.682, correct to 3 places.

Using the Numeric Solver:

Press **[HOME][F2:Algebra][1:Solve]**

Enter **[x][^] 3 [+][x][+] 1[=] 0 [,][x][ )]**

The entry/answer pair displays x=-.682328

To find all solutions use **[F2:Algebra][A:Complex][▷][1:cSolve]**

The complex answers $x \approx .341164 \pm 1.16154i$ are also displayed.

Finding the Intersection of 2 graphs:

ex. Find the intersections of $y = x^2 + 2x - 3$ and $y = \frac{x}{2}$

Press **[GRAPH][Y=]** in empty spaces enter,

*Y*1 =**[x][^] 2 [+] 2 [x] [–] 3**

*Y*2 =**[x][÷] 2**

[F2:Zoom][4:ZoomDec]

You can see the intersections are in the first and third quadrants. **[F3:Trace]** to the intersection in Quad. 1. Using [△] and [▽] arrows you can see that the intersections are not "nice" numbers, so use an

INTERSECTION SOLVER

Press **[F5:Math][5:Intersection][ENTER][ENTER]**

Move the cursor to the left of the intersection,

Press **[ENTER]**

Move the cursor to the right of the intersection

Press **[ENTER]**

The display reads *xc*:1.13746$_E$0 *yc*:5.6873$_E$ –1

The intersection is approximately $x \approx 1.137\ldots \quad y \approx .569\ldots$

Find the intersection in Quad. 3 using the same method.

$x \approx -2.367\ldots \quad y \approx -1.319\ldots$

13. Finding Maximum and Minimum values of a Function:

Find the minimum value of $f(x) = x^2 + 3x - 1$

Using Graph Solve:

Press **[Y=]**, delete or turn off any function entries.

In an empty space enter **[x][^] 2 [+] 3 [x][–]1**

[F2:Zoom][4:ZoomDec].

[F5:Math][3:Minimum]

Move the cursor to the left of the minimum, press **[ENTER]**

Then to the right of the minimum, press **[ENTER]**
On the screen read:
Minimum
xc: -1.5 *yc*: -3.25
The y value $y = -3.25$ is the minimum

Using the Solver:

Press **[F3:Calc][6:fMin(**] In the parenthesis put (expression, variable)
The Entry line is **fmin**$(x^2 + 3x - 1, x)$ **[ENTER]**
The entry/answer pair displays x = - 3/2
To get a decimal answer press [◇]**[ENTER]** x = -1.5
To find the minimum value evaluate $f(1.5)$.
-1.5 **[STO▸]** $x : x^2 + 3x - 1$ **[ENTER]**
read the value y = -3.25

14. Statistic Plots:

Setting up:

In the MODE Menu, select Function
[MODE] [▹] **1 [ENTER]**
In the Y = Menu, delete Y1
[◇][**Y=**], move the cursor to Y1, **[CLEAR]**

Entering Data:

Display the Data/Matrix Editor and create a new data set.
[APPS][6:Data/Matrix Editor][3:New][▽][▽]
Name this set DABC
[2nd][a-lock] D A B C [alpha] [ENTER] [ENTER]

Use the following data

x	1	2	3	4	5	6	7	8
y	1.1	2.6	3.8	5.1	5.9	7.2	8.2	9

Enter the independent variables, x, in list c1. Highlight the first position in c1. Press **1 [ENTER] 2 [ENTER] 3 [ENTER]**...
Enter the dependent variables, y, in list c2. Highlight the first position in c2. Use [◇][△] to move to the top of c2.
Press **1.1 [ENTER] 2.6 [ENTER] 3.8 [ENTER]**...

Plot the data:

Press **[F2:Plot Setup]**
Define the plot parameters. **[F1:Define]**

Set the options:

Plot type = Scatter [▽]
Mark = Box [▽]
x = c1 **[alpha] C 1** [▽]
y = c2 **[alpha] C 2** [▽]
[ENTER][◇]**[Y1**=]

Highlight **Plot 1** [△], press **[F2:Zoom][9:ZoomData]**

Calculate the Regression Analysis:

Display the Calculate dialog box:

[**APPS**][**6**:**Data/matrix Editor**][**1**:**Current**]
[**F5**:**Calc**] Set the following options:
Calculation Type = LinReg
[▷][**5**:**LinReg**][▽]
x = c1 **[alpha] C 1** [▽]
y = c2 **[alpha] C 2** [▽]
Store RegEQ to [▷][▽]
[▽][▽][**ENTER**]

The regression line data are displayed.

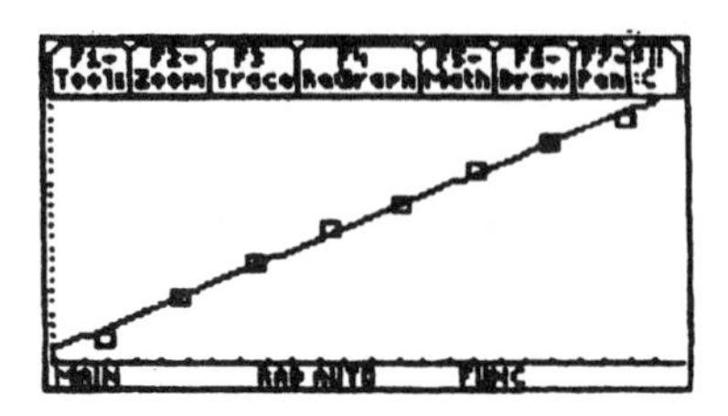

Regression Plot:

[ENTER][◇]**[Y1**=]

Select Y1, press **[F2:Zoom][9:ZoomData]**

The Regression Line is drawn through the Scatter Plot.

15. Complex Numbers:

Rectangular Complex

The number $i = \sqrt{-1}$ is **[2nd][i]** located on the bottom row of the keyboard.

Set the MODE to **Radian**

Mode:

In the MODE menu, set Complex Format as

REAL: will not display a complex number unless you enter a complex number. Display is $a + bi$ or $re^{i\theta}$

RECTANGULAR: Display is $a + bi$

POLAR: Display is $re^{i\theta}$ if angle MODE is Radian
Display is $(r\angle\theta)$ if angle MODE is Degree

To enter 2 + 3i, press **2** [+] **3** **[2nd][i]**

To multiply or divide complex numbers, each number must be entered in a parenthesis.

ex. (**2** + **3i**) + (**3** – **5i**) = $\underline{5 - 2i}$

Polar Complex:

The number a + bi is written $re^{i\theta}$

$r = \sqrt{a^2 + b^2} \quad \theta = \arctan(\frac{y}{x})$

Complex Functions:

Press **[2nd][MATH][5:Complex]** to display the list of functions and operations for complex numbers

16. **Matrices**:

Matrices are *stored* by name, edited and used in matrix arithmetic from the keyboard, or in the APPS Data/Matrix Editor

Matrices are stored in the APPS Data/Matrix Editor.

They may also be entered and named from the keyboard.

Enter [[ ][[ ] **1**, **2** [ ]][[ ] **3** , **4** [ ]][ ]][**ENTER][STO▸]**

[2nd][QUIT] the matrix [A] = $\begin{bmatrix} 1 & 2 \\ 3 & 4 \end{bmatrix}$ is saved.

Matrix *operations* are displayed with **[MATRX]**[▹] (MATH)

From this screen press **[ALPHA] B**

The home screen displayes **rref(**

Press **[MATRX][1**:[A]][)][**ENTER]**

Matrix [A] is row reduced to $\begin{bmatrix} 1 & 0 \\ 0 & 1 \end{bmatrix}$

Matrix *arithmetic* +, –, ×,^ is done with the arithmetic keys on the keyboard.

Matrix *Row Operations* are done from the menu

[2nd][MATH][4:Matrix][J:Row ops▸]

Press [▹]. The row operations are **1, 2, 3, 4.**

See the TI89 manual for details.

Find a Solution to a Linear System:

ex. Solve the system of equations:

$-2x_1 + 3x_2 - 2x_3 = 16$

$-5x_1 + 3x_2 - 5x_3 = 22$

$x_1 \quad + x_3 = -2$

Enter the matrix $\begin{bmatrix} -2 & 3 & -2 & 16 \\ -5 & 3 & -5 & 22 \\ 1 & 0 & 1 & -2 \end{bmatrix}$ **[STO▸]** [⇑] **A**

Use [(–)] for the negative numbers

[2nd][[ ][2nd][[ ] -2 , 3 , -2 , 16 [2nd][ ]][2nd][[ ] -5 , 3 , -5 , 22
[2nd][ ]][2nd][[ ] 1 , 0 , 1 , -2 [2nd][ ]][2nd][ ]][ENTER]

[STO▸][⇑][A][ENTER]

Find the reduced row echelon form:

[2nd][MATH][4:Matrix][4:rref(][⇑][A][)][ENTER]

$x = 2 - \alpha, \quad y = 4, \quad z = \alpha.$

ex.

Enter and store the matrix: $\begin{bmatrix} 1 & 1 & 2 & 1 \\ -1 & 1 & 0 & 1 \\ 2 & 1 & 1 & 0 \\ 1 & 3 & 1 & 0 \end{bmatrix}$

Use [(–)] for a negative number

[2nd][[ ][2nd][[ ]1 , 1 , 2 , 1 [2nd][ ]]

[2nd][[ ] -1 , 1 , 0 , 1[2nd][ ]]

[2nd][[ ] 2 , 1 , 1 , 0 [2nd][ ]]

[2nd][[ ] 1 , 3 , 1 , 0 [2nd][ ]]

[2nd][ ]][ENTER][STO▸][⇑][A][ENTER]

Find and store the inverse:

[⇑][A][^][-1][ENTER][STO▸][⇑][B][ENTER]

To check the inverse:

[⇑][A][×][⇑][B][ENTER]

The answer is the Identity of order 4.

To find the determinent of matrix A:

[2nd][MATH][4:Matrix][2:det(][⇑][A][)][ENTER]

The answer is 6

17. Algebra

The TI-89 calculator is really a mini-computer. It contains a CAS (Computer Algebra System) which allows it to do algebraic computations as well as arithmetic calculations.

Some Algebra Examples:

Use **[MENU] A** (Algebra)

ex. expand $(x+y)^4$

Press **[F2:Algebra]**

[3:expand(]

(x + y)^4 [ENTER]

$x^4 + 4x^3y + 6x^2y^2 + 4xy^3 + y^4$

ex. factor $(x^3 - y^3)$

Press **[F2:Algebra]**

[2:factor]

(x^3 – y^3) **[ENTER]**
$(x-y)(x^2+xy+y^2)$

ex. Add the fractions $\frac{x+1}{x-2}+\frac{x}{x+3}$
Press [**F2**:**Algebra**]
[**6**:**comDenom(**] Enter
$(x+1)\div(x-2)+x\div(x+3))$
[ENTER]
$\frac{2x^2+2x+3}{x^2+x-6}$.

Chapter 6 – Using the TI-92 Graphing Calculator.

1. Notation:

Keystrokes, except for numbers are in bold brackets **[MATH]**.

For a second function √‾, press **[2nd]** [√‾], the second functions are printed in yellow on the left over the keys.

There are two **[2nd]** keys, they both do the same thing.

Letters are typed from the QWERTY keyboard. The default is lower case.

To enter **A** press [⇑] **A**, to set the cap-lock, press **[2nd][CAPS]**

For functions in the screen menus ≤ press **[2nd]** **[MATH][8:Test]** [**3**:≤]. To execute a screen menu function say Test, you may press the number **8**, or use [▽] to move down and highlight the entry [**8**:**Test**] and press [**ENTER**].

The green diamond key [◇] is used for functions on the keyboard that are in green on the right over some keys. It also is used with other keys to activate menus and applications.

The shift key [↑] is also used for highlighting in editing.

Calculator results are underlined.

2. Basics:

To adjust the screen contrast

Press and hold both [◇] and [+] to increase the contrast (darken).

Press and hold both [◇] and [–] to decrease the contrast (lighten).

Entry/Answer pairs in the history area are displayed in "pretty print".

Mode settings: Modes control how numbers and graphs are displayed and interpreted.

Press**[MODE]**, select the following settings. Leave these settings unless directed to change a mode setting.

Graph	Function
Display Digits	Float 6
Current Folder	Main
Angle	Radian
Exponential Format	Normal
Complex Format	Real
Vector Format	Rectangular
Pretty Print	On
Split Screen	Full
Split 1 App	Home

Press **[ENTER]** to leave the MENU
(Auto gives exact answers if possible, otherwise a decimal)

3. Arithmetic Computations:

Order of Operations: The order is Algebraic, operations are performed from left to right. First exponentiation, then multiplication and division, then addition and subtraction.

Parenthesis must be used to change the algebraic order.

Minus sign: Use [(–)] for negative, [–] for subtraction.

Fractions: to enter fractions use [÷],
To change a fraction to a decimal use [◇][**ENTER**]
To change back to a fraction use **[ENTER]**

To calculate: press **[ENTER]**

ex. $2 \times 3 + 4 \times 5$ **[ENTER]** 26
$2 \times (3 + 4) \times 5$ **[ENTER]** 70
$3 + 4 \div 2$ **[ENTER]** 5
$(3 + 4) \div 2$ **[ENTER]** 7/2

The answer is in fractional form because the "Pretty Print" mode is on.
To get a decimal answer press [◇][**ENTER**] 3.5

4. Scientific Notation:

From the keyboard type the number between 1 and 10 **[2nd][EE]** exponent.
The exponent must be an integer between -99 and 99.
To display answers in Scientific Notation change the MODE menu entry.
[MODE] use [▽] to Exponential Format [▷] **[2:SCIENTIFIC]**
[ENTER]
You may want to change the number of decimal places displayed. To display 3 places, change to FIX 3 in the MODE menu.
To return to Normal
[MODE] use [▽] to Exponential Format [▷] **[1:NORMAL]**
[ENTER]

5. Editing (on the Entry Line):

The TI-89 has both an insert and overtype mode. Insert is default.
To change between insert and overtype press **[2nd][INS]**.

To move the cursor
To the beginning of an expression press **[2nd]**[◁].
To the end of an expression press **[2nd]**[▷].

To Delete

The symbol to the left of the cursor press [←]

The symbol to the right of the cursor press [◇][←]

All symbols to the right of the cursor press **[CLEAR]**

The entire line if the cursor is at the beginning or the end of the entry line press **[CLEAR]**

The entire line if the cursor is in the middle of the entry line then press **[CLEAR][CLEAR]**

To highlight Multiple Characters

Move the cursor to the left of the symbols.

Hold [↑] and press [▷] for each symbol to be highlighted.

Retype the symbols or press [←] to delete them.

To remove a highlight use [▷] or [◁] to move over the highlighted area.

To highlight the Entry Line, press **[ENTER]**

To edit an entry in the history area, press **[2nd][ENTRY]** repeatedly until it moves back to the entry line.

To clear the history area press **[F1][8:Clear Home]**

To exit a MENU press **[ESC]** or **[2nd][QUIT]**

To erase a function from the Y= screen press **[CLEAR]**

To clear a history pair, highlight the entry, press **[CLEAR]**

To return to the Home Screen, press **[HOME]**

6. Entering Algebraic Expressions:

In the Y = Menu

To display the Y = Menu press [◇][Y=]

Enter the function. It is now available to use as a graph or table.

This menu holds 99 items.

On the Homescreen, enter the function. When you press **[ENTER]** it will be evaluated with the values of the variables that are stored in Memory.

The Catalog is a list of all functions, operations and symbols in the calculator.

Press **[CATALOG]**, you will see a list sorted alphabetically. Press the first letter of the word you want, use the down arrow to move the pointer to the entry, press **[ENTER]** and the word is pasted onto the homescreen.

The symbols are listed after Z, use [△] from A.

When you enter built in functions, left parenthesis are included. Complete the input variable, then type the right parenthesis.

Special Functions:

To change exact answers to decimal, [◇][**ENTER**]

Absolute value of x: press **[2nd][MATH][1:Number]**[▷]**[2:abs][x]**

Powers, x^5: press **[x]**[^] **5**

2^x: press **2** [^]**[x]**

e^x: press [◇]**[x]**[)] the ^and (are forced.

Roots, $\sqrt[5]{x}$: press **[x]**[^][(]**1** [÷] **5**[)]

Natural log, $\ln(x+5)$: press **[2nd]**[ln]**[x]**[+] **5**

Trig function, $\tan(3x^2)$: press **[tan] 3 [x]**[^] **2**

Inverse trig function, arcsine(x): [◇][**sin**$^{-1}$]**[x]**[)]

Factorial, 5!: press **5 [MATH][7:Probability][ENTER][1:!]**

Combinations, ${}_5C_2$: press **5 [MATH][7:Probability][ENTER]** **[3:**${}_nC_r$**] 2**

7. **Evaluating a function**: (**checking your answer**)

Numbers can be stored as variables. A variable name can have 1-8 symbols (letters or numbers), the first one can not be a number.

Numbers are stored in these positions using the command **[STO▸]**, an arrow → is shown on the homescreen.

Single letter Variables will be used in this text.

When an algebraic expression is entered, the variables are replaced with the stored constants and a number is displayed.

To evaluate a function on the home screen use the key **[2nd]**[:]to connect commands.

ex. To evaluate $x^2 + 3x - 1$ for $x = 2$,
Press **2 [STO▸]**[**x**][**2nd**][:]
[x][x²][+] **3 [x]**[–] **1 [ENTER]**

the result is 9.

All expressions containing an x will be evaluated with x = 2 until this number in the x box is replaced with another number.

To clear single variables press **[F6][1:Clr a-z]**

8. **Angles**:

The type of measure for angles is set in the MODE menu.

Set **[MODE]** to either **Radian** or **Degree** by highlighting Angle and pressing [▷]**[1:RADIAN]** or **[2:DEGREE][ENTER]**

After the MODE is set, use the operations in **[2nd][MATH]** **[2:Angle]** to change the units.

To enter an angle in Degrees while in **Radian** MODE:

ex. Press **[SIN] 30 [2nd][MATH][2:Angle][1:°][)][ENTER]** .5

To enter an angle in Radians while in **Degree** MODE:

ex. Press **[SIN]** $\pi \div 6$ **[2nd][MATH][2:Angle][2:ʳ][)][ENTER]** .5

To change Rectangular Coordinates to Polar (Degree Mode):

ex. Press **[2nd][MATH][2:Angle][5:R▸Pr(] 1,1 [)][ENTER]**

1.4142... is the radius in polar form.

Press **[2nd][MATH][2:Angle][6:R▸Pθ(] 1,1 [)][ENTER]**..

45 is the angle in **Degrees.**

To change Polar Coordinates (Degree Mode) to Rectangular:

ex. Press **[2nd][MATH][2:Angle][3:P▸Rx(]** $\sqrt{2}$**,45 [)][ENTER]** 1

is the x coordinate in rectangular coordinates.

Press **[2nd][MATH][2:Angle][4:P▸Ry(]** $\sqrt{2}$**,45 [)][ENTER]**....1

If the MODE is Radians, enter the angle in Radians.

[HOME] returns to Home Screen.

9. **Building Tables**:

The TI-89 has a built in table function used to evaluate functions for different values of x. The algebraic expression to be evaluated is stored in the Y = menu. In the TABLE SET menu,
set the initial value for x and the
step sixe between entries, or set ASK to enter a random
set of numbers for x.

ex. Enter $x^2 + 3x - 1$ in Y1 in the Y = menu.
Press **[APPS][5:Table][F2:Set up]**
set the following values
[(–)] **4 [ENTER]** for TblStart
1 [ENTER] for ∆Tbl
Off for Graph <-> Table
Auto for Independent
Press **[ENTER]** when finished.

Use the up and down arrows to see more values in the table.
To return to the Home Screen, press **[HOME]**
Tables may also be constructed in the STAT menu to examine and evaluate statistical data .

10. **Building Lists**:

Numbers can be stored in Lists either on the home screen or in the Data/Matrix Editor menu.
Lists may be named (stored) by naming them.

ex. From the Home Screen,enter the numbers 1, 3, 5, 7 in list **L1**
[2nd][{] 1 , 3 , 5 , 7 [2nd][}][STO▸][↑] L 1 [ENTER]
Lists from the keyboard are stored in the Data/Matrix Editor menu.
To display L1 on the home screen, press [↑] **L 1 [ENTER]**
Arithmetic operations are done using the +,–,×,÷,^ keys,
when the lists are the same length.
Other operations are done from the **[2nd][MATH][3:List]** menu.

Special Lists:
[2nd][MATH][3:List][1:seq(] expression, variable, begin, end increment)
Generate a sequence of numbers using a rule for expression.
ex. **[2nd][MATH][3:List]**
[1:seq(][x][^] 2, [x] , 1 , 5 , 2][)][ENTER]
square the numbers starting with 1, adding 2, ending at 5.
The result is { 1, 9, 25 }
[2nd][MATH][3:List][7:cumSum(] (list) returns a list of cumulative sums starting with the first entry.
ex. For L1, { 1 , 4 , 9 , 16 }
[2nd][MATH][3:List][6:sum(] (list) returns the sum of all the numbers in the list.
ex. For L1, the sum is 16.

11. Graphing Functions:

The function to be graphed must be entered as one of the 99 functions in the Y = menu. Enter $x^2 + 3x - 1$ in Y1.
The function must be turned on to graph. The ✓ must be highlighted.
To turn the function on or off, highlight the function and press **[F4]**.
Each time **[F4]** is pressed ✓ reverses the current state.
To graph the function press [◇] **[GRAPH]**
The functions with a ✓ are graphed on the current window settings.
The ZOOM menu **[F2:Zoom]** has choices to zoom in or out and for built in window sizes.
Press **[F2:Zoom]**, press **[4:ZoomDec]** The window is set to a Friendly window. Each pixel is 0.1 units, and the graph is its true shape. Unit size on X axis = unit size on Y axis. Any multiple of these settings also makes a Friendly window.
Press **[ZOOM][6:ZoomStd]** This is the Standard window, it sets X to [-10,10] and Y to [-10,10]. The units on the X and Y axis are not the same size, so graphs are not their true shape.
Press **[ZOOM][5:ZSquare]** Now the graph is its true shape again.
The window settings for the Y axis are left the same, and the X settings have been recalculated so that the units on the X and Y axis are the

same. This is called a Square window.

Press [◇] [**WINDOW**] It displays the current settings for the X and Y windows. These values can be changed by using the arrow keys to highlight the entry, pressing [**CLEAR**] then typing the new value. Use [**ENTER**] after each change. (Remember to use (–) for a negative number). Press [◇] [**GRAPH**] to return to the graph.

Press [**F3:TRACE**] A blinking star cursor appears on the Y axis. The left and right arrow keys move this cursor along the function graph. The coordinates that are displayed are points that satisfy the function equation. The Y values are computed from the X value of the pixel. When more than one function is plotted, the up and down arrows move the cursor vertically between the different plots. These coordinates are only approximations unless the graph is on a Friendly window and the coordinate is a rational number.

To graph a *split function* $f(x) = \left\{ \begin{array}{ll} x+7, & x \le -5 \\ -x+4, & x > -5 \end{array} \right\}$

Press [**Y=**] Then enter the function in an empty space.
[**2nd**][**alpha**][**w**][**h**][**e**][**n**][**alpha**][**(**] $x \le -5, x+7, -x+4)$
The symbol [≤] is in [**CATALOG**]
The vertical line between the segments is a technology error.

To graph *Parametric Equations* press [**MODE**] from **Function** use [▷] to highlight **Parametric**, then press [**ENTER**]
Press[◇] [**Y=**] the menu now reads $\mathbf{X}_{1T}$= and $\mathbf{Y}_{1T}$=
You may want to change the window settings. Now X and Y are both functions of the independent variable T.
The settings for Tmin and Tmax determine how much of the graph is plotted.
Tstep determines how many points are plotted.
Enter the functions and press [**GRAPH**]

ex. Write the function $y = x^2 + 3x - 1$ in parametric form.
Let $x = t, y = t^2 + 3t - 1.$
Enter $\mathbf{X}_{1T} = t$ and $\mathbf{Y}_{1T} = t^2 + 3t - 1$
Press [**WINDOW**] Set **Tmin** = (-)5, **Tmax** = **5**.
Press [◇][**GRAPH**] The graph is displayed.

To graph *implicit functions* easily, write the equation in Parametric form.

ex. Write the equation $x^2 + y^2 = 4$ in parametric form:
$x = 2\cos(t), y = 2\sin(t).$
The graph is a circle.
The circle is not round, press [**F2**][**6:ZoomStd**]

The regraphed circle is almost round.

Remember to change back to Function Mode.

To graph *Polar Equations*: press **[MODE] Function**

Use [▷][**3:POLAR**] press **[ENTER]**

Press [◇][**Y=**] Enter **r1**=θ θ is on the keyboard.

[F2:Zoom][4:ZoomDec]

Change θmax. to 6π.

[F2:Zoom][4:ZoomDec]

Remember to change back to Function Mode.

12. Solving Equations

Write the equation in the form $f(x) = 0$

Using Trace and Zoom will give you an approximate answer. Use either a Zoom box or use Zoom In.

ex. Find the real zero of $x^3 + x + 1 = 0$ correct to hundredths.

Use Zoom In:

Press [◇] [**Y=**]

move to Y1, press **[CLEAR]**

Enter **[x][^] 3** [+]**[x]**[+]**1**

[F2:Zoom][4:ZoomDec]

You can see that the function crosses the X axis between -1 and 0.

Press **[F3:Trace]**, move the cursor near the root.

[F2:Zoom][2:Zoom In][ENTER] Repeat

[F3:Trace], move the cursor near the root.

[F2:Zoom][2:Zoom In][ENTER] Repeat

[F3:TraceE], move the cursor near the root.

[F2:Zoom][2:Zoom In][ENTER]

[Trace]

For x = -.682813, y = -.001162

For x = -.679688, y = .006314

The value of the zero is x ≈ -.68 correct to hundreths.

Note that these 2 values of y lay on opposite sides of the X axis.

You may not get exactly the same numbers.

(These figures depend on the value of the Zoom Factors, these are set at 4).

Use Zoom box:

Press[◇] **[Y=]** in Y1, **[CLEAR]** then enter **[x][^] 3** [+] **[x]**[+]**1** **[F2:Zoom][4:ZoomDec]**

You can see that the function crosses the X axis between -1 and 0.

Move the free cursor with the arrow keys until it is above and to the left of the zero. Press **[F2:Zoom][1:ZoomBox][ENTER]**
Note that the cursor is now a blinking box.
Use [▷], then [▽] to draw a box with the zero inside.
Repeat until the tick marks on the X axis are clearly visible.
Trace to approximate the value x ≈ -.68

Using the Graph solve feature:

ex. Find the real zero of $x^3 + x + 1 = 0$ correct to hundreths.
Press[◇] **[Y=]** in an empty space enter **[x]**[^] **3** [+] **[x]**[+]**1** **[F2:Zoom][4:ZoomDec]**.
You can see that the function crosses the X axis between -1 and 0.
Press **[F5:Math][2:Zero]**
The graph is drawn with the request *Lower Bound?*
Move the blinking cursor with [◁] until it is to the left of the zero, **[ENTER]**
Now the request is *Upper Bound?*
Use [▷] to move the cursor to the right of the zero, **[ENTER]**
On the bottom of the screen, read
Zero
xc: -.682328 *yc*: 0

The approximation is x ≈ -.682, correct to 3 places.

Using the Numeric Solver:

Press **[HOME][CATALOG]**
Use [▽] until the ▸ points to *solve*.
Press **[ENTER]** to paste the command on the Entry Line.
Enter **[x]**[^] **3** [+]**[x]**[+] **1** [=] **0** [,]**[x]**[ )]
The entry/answer pair displays x ≈ -.682328

Finding the Intersection of 2 graphs:

ex. Find the intersections of $y = x^2 + 2x - 3$ and $y = \frac{x}{2}$
Press **[GRAPH][Y=]** in empty spaces enter,
$Y1$ =**[x]**[^] **2** [+] **2** **[x]** [–] **3**
$Y2$ =**[x]**[÷] **2**
[F2:Zoom][4:ZoomDec]
You can see the intersections are in the first and third quadrants. **[F3:Trace]** to the intersection in Quad. 1. Using [△] and [▽] arrows you can see that the intersections are not "nice" numbers, so use an
INTERSECTION SOLVER
Press **[F5:Math][5:Intersection][ENTER][ENTER]**

Move the cursor to the left of the intersection,
Press **[ENTER]**
Move the cursor to the right of the intersection
Press **[ENTER]**
The display reads *xc*:1.13746 *yc*: 0.56873
The intersection is approximately $x \approx 1.137...$ $y \approx .569...$
Find the intersection in Quad. 3 using the same method.
$x \approx -2.367...$ $y \approx -1.319...$

13. Finding Maximum and Minimum values of a Function:

Find the minimum value of $f(x) = x^2 + 3x - 1$
Using Graph Solve:
Press [**Y**=], delete or turn off any function entries.
In an empty space enter **[x]**[^] **2** [+] **3** **[x]**[–]**1**
[F2:Zoom][4:ZoomDec].
[F5:Math][3:Minimum]
Move the cursor to the left of the minimum, press **[ENTER]**
Then to the right of the minimum, press **[ENTER]**
On the screen read:
Minimum
xc: -1.5 *yc*: -3.25
The y value y = -3.25 is the minimum

Using the Solver:

Press [**F3**:**Calc**][**6**:**fMin**( ] In the parenthesis
put (expression, variable)
The Entry line is **fmin**($x^2 + 3x - 1, x$) **[ENTER]**
The entry/answer pair displays x = - 3/2
To get a decimal answer press [◇]**[ENTER]** x = -1.5
To find the minimum value evaluate $f(1.5)$.
-1.5 **[STO▸]** x [:] $x^2 + 3x - 1$ **[ENTER]**
read the value y = -3.25

14. Statistic Plots:

Setting up:

In the MODE Menu, select Function
[MODE] [▹] **1** **[ENTER]**
In the Y = Menu, delete Y1
[◇][**Y**=], move the cursor to Y1, **[CLEAR]**

Entering Data:

Display the Data/Matrix Editor and create a new data set.
 [APPS][6:Data/Matrix Editor][3:New][▽][▽]
Name this set DABC
 [2nd][a-lock] D A B C [alpha] [ENTER] [ENTER]

Use the following data

x	1	2	3	4	5	6	7	8
y	1.1	2.6	3.8	5.1	5.9	7.2	8.2	9

Enter the independent variables, x, in list c1. Highlight the first position
 in c1. Press **1 [ENTER] 2 [ENTER] 3 [ENTER]...**
Enter the dependent variables, y, in list c2. Highlight the first position
 in c2. Use [◇][△] to move to the top of c2.
 Press **1.1 [ENTER] 2.6 [ENTER] 3.8 [ENTER]...**

Plot the data:

Press **[F2:Plot Setup]**
Define the plot parameters. **[F1:Define]**
Set the options:

Plot type	=	Scatter [▽]	
Mark	=	Box	[▽]
x	=	c1	**[alpha] C 1** [▽]
y	=	c2	**[alpha] C 2** [▽]

 [ENTER][◇]**[Y1=]**
Highlight Plot 1 [△], press **[F2:Zoom][9:ZoomData]**

Calculate the regression Analysis:

Display the Calculate dialog box:
 [APPS][6:Data/matrix Editor][1:Current]
 [F5:Calc] Set the following options:
 Calculation Type = LinReg
 [▷]**[5:LinReg]**[▽]

x	=	c1	**[alpha] C 1** [▽]
y	=	c2	**[alpha] C 2** [▽]
Store RegEQ to			[▷][▽]

 [▽][▽]**[ENTER]**
The regression line data are displayed.

Regression Plot:

 [ENTER][◇]**[Y1=]**
Select Y1, press **[F2:Zoom][9:ZoomData]**
The Regression Line is drawn through the Scatter Plot.

15. Complex Numbers:

Rectangular Complex:

The number $i = \sqrt{-1}$ is **[2nd][i]** located on the bottom row of the keyboard.

Set the MODE to **Radian**

Mode:

In the MODE menu, set Complex Format as

REAL: will not display a complex number unless you enter a complex number. Display is $a + bi$ or $re^{i\theta}$

RECTANGULAR: Display is $a + bi$

POLAR: Display is $re^{i\theta}$ if angle MODE is Radian

Display is $(r\angle\theta)$ if angle MODE is Degree

To enter 2 + 3i, press **2** [+] **3** **[2nd][i]**

To multiply or divide complex numbers, each number must be entered in a parenthesis.

ex. **(2 + 3i)** + **(3 – 5i)** = $\underline{5 - 2i}$

Polar Complex:

The number a + bi is written $re^{i\theta}$

$r = \sqrt{a^2 + b^2} \quad \theta = \arctan(\frac{y}{x})$

Complex Functions:

Press **[2nd][MATH][5:Complex]** to display the list of functions and operations for complex numbers

16. Matrices:

Matrices are <u>*stored*</u> by name, edited and used in matrix arithmetic from the keyboard, or in the APPS Data/Matrix Editor

Matrices are stored in the APPS Data/Matrix Editor.

They may also be entered and named from the keyboard.

Enter [[ ][[ ] **1**, **2** [ ]][[ ] **3** , **4** [ ]][ ]][**ENTER][STO▸]**

[2nd][QUIT] the matrix [A] = $\begin{bmatrix} 1 & 2 \\ 3 & 4 \end{bmatrix}$ is saved.

Matrix <u>*operations*</u> are displayed with **[MATRX]**[▹] (MATH)

From this screen press **[ALPHA] B**

The home screen displayes **rref(**

Press **[MATRX][1**:[A]][)]**[ENTER]**

Matrix [A] is row reduced to $\begin{bmatrix} 1 & 0 \\ 0 & 1 \end{bmatrix}$

Matrix <u>*arithmetic*</u> +, –, ×,^ is done with the blue operation keys on the keyboard.

Matrix *Row Opreations* are done from the menu

[2nd][MATH][4:Matrix][J:Row ops▸]

Press [▹]. The row operations are **1, 2, 3, 4.**

See the TI92 manual for details.

Find a Solution to a Linear System:

ex. Solve the system of equations:

$$-2x_1 + 3x_2 - 2x_3 = 16$$
$$-5x_1 + 3x_2 - 5x_3 = 22$$
$$x_1 \qquad + x_3 = -2$$

Enter the matrix $\begin{bmatrix} -2 & 3 & -2 & 16 \\ -5 & 3 & -5 & 22 \\ 1 & 0 & 1 & -2 \end{bmatrix}$ **[STO▸] [⇑] A**

Use [(–)] for the negative numbers

[2nd][[ ][2nd][[ ] -2 , 3 , -2 , 16 [2nd][ ]][2nd][[ ] -5 , 3 , -5 , 22
[2nd][ ]][2nd][[ ] 1 , 0 , 1 , -2 [2nd][ ]][2nd][ ]][ENTER]
[STO▸][⇑][A][ENTER]

Find the reduced row echelon form:

[2nd][MATH][4:Matrix][4:rref(][⇑][A][)][ENTER]

$x = -(2 - \alpha) \quad y = 4 \quad z = \alpha$

ex. *Find the inverse of*

Enter and store the matrix: $\begin{bmatrix} 1 & 1 & 2 & 1 \\ -1 & 1 & 0 & 1 \\ 2 & 1 & 1 & 0 \\ 1 & 3 & 1 & 0 \end{bmatrix}$

Use [(–)] for a negative number

[2nd][[ ][2nd][[ ]1 , 1 , 2 , 1 [2nd][ ]]
[2nd][[ ] -1 , 1 , 0 , 1[2nd][ ]]
[2nd][[ ] 2 , 1 , 1 , 0 [2nd][ ]]
[2nd][[ ] 1 , 3 , 1 , 0 [2nd][ ]]
[2nd][ ]][ENTER][STO▸][⇑][A][ENTER]

Find and store the inverse:

[⇑][A][^][-1][ENTER][STO▸][⇑][B][ENTER]

To check the inverse:

[⇑][A][×][⇑][B][ENTER]

The answer is the Identity of order 4.

To find the determinent of matrix A:

[2nd][MATH][4:Matrix][2:det(][⇑][A][)][ENTER]

The answer is 6

17. Algebra:

The TI-92 calculator is really a mini-computer. It contains a CAS (Computer Algebra System) which allows it to do algebraic computations as well as arithmetic calculations.

Some Algebra Examples:

Use **[MENU] A** (Algebra)

ex. expand $(x+y)^4$

Press **[F2:Algebra]**

[3:expand(]

(x + y)^4 [ENTER]

$x^4+4x^3y+6x^2y^2+4xy^3+y^4$

ex. factor (x^3-y^3)

Press **[F2:Algebra]**

[2:factor]

(x^3 – y^3) [ENTER]

$(x-y)(x^2+xy+y^2)$

ex. Add the fractions $\frac{x+1}{x-2}+\frac{x}{x+3}$

Press **[F2:Algebra]**

[6:comDenom(] Enter

$(x+1)\div(x-2)+x\div(x+3))$

[ENTER]

$\frac{2x^2+2x+3}{x^2+x-6}$

Chapter 7
Using the Casio CFX-9850 G Plus and the Casio CFX-9850 GB Plus

1. Notation:

Keystrokes, except for numbers, letters and commas are in bold brackets **[MATH]**
For a second function $\sqrt{\ }$, press **[SHIFT]** [$\sqrt{\ }$] the second functions are
printed in yellow on the left over the keys.
For the letter A, press **[ALPHA][A]** the ALPHA symbols are printed in
red on the right over the keys.
For functions in the screen menus, 2! from the RUN menu, press
[OPTN][F6:▷][F3:PROB] 2 [F1:*x*!][EXE]
In this text 2! is keystroked **2 [F1:*x*!]**
Calculator results are underlined.

2. Basics:

Casio is a MENU driven calculator.
The MAIN MENU screen contains ICONS to set
the MODE for your work.
Highlight the ICON and press **[EXE]** or
press the number or letter printed in
the lower right corner of the ICON.

MAIN MENU
RUN STAT MAT LIST GRAPH
DYNA TABLE RECUR CONICS EQUA
PRGM TVM LINK CONT MEM

To adjust the screen contrast: or adjust the color
Press **[MENU] E** (on 9850G press **D**)
Tag ▸ the line you wish to change
with the [△] and [▽] arrow keys, then make
changes with the [▷] and [◁] arrow keys.
Mode settings: Enter the MENU you are going to use,
Press **[SHIFT][SET UP]** When you highlight a MODE setting,
the choices are in the screen menu.
To change a mode, press the green function key
below your choice on the screen menu.

Use the following choices from the RUN menu
unless the directions say otherwise.

Mode	:Comp
Func Type	:Y=
Draw Type	:Connect
Derivative	:Off
Angle	:Rad
Coord	:On
Grid	:Off
Axes	:On
Label	:On
Display	:Normal

3. **Arithmetic Computations**:

Press **[MENU] 1** (RUN)

Order of Operations: The order is Algebraic, operations are performed from left to right. First exponentiation, then multiplication and division, then addition and subtraction. Parenthesis must be used to change the algebraic order. Parenthesis must be used around the numerator and denominator in Algebraic Fractions.

Minus sign: Use [(–)] for negative, [–] for subtraction.

Fractions: to enter fractions use [$a^b/_c$] the answer is
a mixed number displayed 1⌟2⌟3 for $1\frac{2}{3}$.

To change to an improper fraction press **[SHIFT]**[$^d/_c$]
$\frac{5}{3}$ is displayed 5⌟3

To change a fraction to a decimal use **[F↔D]**

To change a decimal to a fraction use **[F↔D]** provided
the numbers were entered as fractions.

The Casio computes using fractions if there was
a fraction in the entry line, otherwise it computes
using decimals.

To calculate: press **[EXE]**

ex.	$2 \times 3 + 4 \times 5$	**[EXE]**	26
	$2 \times (3 + 4) \times 5$	**[EXE]**	70
	$3 + 4 \div 2$	**[EXE]**	5
	$(3 + 4) \div 2$	**[EXE]**	3.5
	(3 + 4)**[a^b/c] 2**	**[EXE]**	3⌟1⌟2

[SHIFT][d/c] 7/2
[F↔D] 3.5
[F↔D] 3⌟1⌟2

4. **Scientific Notation**:

Press **[MENU] 1** (RUN)
From the keyboard use number between 1 and 10 **[SHIFT]**[10^x] exponent **[EXE]**
The exponent must be a number between -99 and 99.
To Display answers in Scientific Notation change the MODE menu entry.
From menu RUN, press **[SHIFT][SET UP]** press [▽] 9 times,
Display is highlighted.
Press **[F2:Sci][F4: 3][EXIT][EXE]**
There are 3 digits displayed.
To return to Normal: Highlight **Display** in the SET UP menu
Press **[F3:Norm][EXIT]**

5. **Editing**:

Before typing **[*EXE*]**

A blinking line determines the current position on the screen.
The arrow keys move the cursor around the screen.

To change an entry move the blinking cursor to the entry, then type the new entry. It will replace the old entry.
To delete a symbol, move to the symbol, press **[DEL]**
To insert a symbol, move to the symbol after the insertion point, press **[SHIFT][INS]** then type the new text.

After typing **[*EXE*]**

Press [▷]to return to the front of the last entry.
Press [◁]to return to the back of the last entry.
Edit as explained in *Before typing EXE*
You may press **[AC/ON]** then [△] to go back to the previous command, continue to press [△] to keep going back to earlier commands.

To clear the home screen press **[AC/ON]**
To exit a MENU, press **[EXIT]** or enter another MENU.
To delete an Error message press **[AC/ON]**
To delete data in MEMORY, press **[MENU][F]** (MEM)(**E** on 9850)
Highlight *Memory Usage* **[EXE]** Highlight the category with the items to delete. Press **[F1:DEL]** The items saved in the category are listed. Press the key for the item to delete, then

[F1:DEL][F1:YES] Press **[MENU]** when done.
To erase a function from the Y = screen, press **[F2:DEL][F1:YES]**

6. **Entering Algebraic Functions:**

Press **[MENU] 5** (GRAPH)
From the Y = screen, use the arrow keys to move to an empty position.
Enter the function. It is now available to use as a graph or a table
This menu holds 20 functions.

Press **[MENU] 1** (RUN)
On the home screen, enter the function. When you press **[EXE]** it will be evaluated with the values of the variables that are stored in Memory.

When you enter built in functions, parenthesis should be used.
Type left parenthesis, complete the input variable,
then type the right parenthesis.
The key **[x,θ,T]** prints x in Function mode, θ in polar mode, and
T in parametric mode.

Special Functions:
From the keyboard:
Powers, x^5: press **[x,θ,T][^] 5**
2^x: press **2** [^]**[x,θ,T]**
e^x: press **[SHIFT]**[e^x]**[x,θ,T]** no ^ is needed.
Roots, $\sqrt[5]{x}$: press **5 [SHIFT]**[$\sqrt[x]{\ }$]**[x,θ,T]**or **[x,θ,T][^][(]1 [a^b/c] 5 [)]**
Natural log, $ln(x+5)$: press [ln][(]**[x,θ,T]**[+] **5** [)]
Conjugate Pairs, $\pm\sqrt{4-x^2}$: press **[SHIFT][{] -1, 1 [SHIFT][}]**
[SHIFT][$\sqrt{\ }$][(] **4**[−]**[x,θ,T]**[x^2][)]
Trig function, $tan(3x^2)$: press **[tan]**[(] **3 [x,θ,T][x²]**[)]
Inverse trig function, $arcsine(x)$: press **[SHIFT][sin⁻¹]**[(]**[x,θ,T]**[)]
From RUN **[OPTN][F6:▷]**
Absolute value of x: press **[F4:NUM][F1:Abs][x,θ,T]**
Factorial, 5!: press **[F3:PROB] 5 [F1:**x!**]**
Combinations, ${}_5C_2$: press **5 [F3:PROB][F3:**${}_5C_2$**] 2 [EXE]**

7. **Evaluating a function: (checking your answer)**

Variables are represented as single alphabetic characters, A through Z, r and θ. Numbers are stored in these positions using [→]
the key above the **[AC/ON]** key.
When an algebraic expression is entered, the variables are replaced with the stored constants and a number is displayed.

To evaluate a function on the home screen use the key **[SHIFT]** [↵] (over EXE) to connect commands.

ex. To evaluate $x^2 + 3x - 1$ for $x = 2$,
From RUN,
Press **[MENU] 1** (RUN)
Press **2 [→][x,,θ,T][SHIFT]**[↵]
[x,θ,T][x²][+] **3 [x,θ,T]**[−] **1 [EXE]**
the result is 9.

If the function is stored in the Y = menu it may be recalled from the VARS menu.

ex. Enter the function $x^2 + 3x - 1$ in the Y = menu as Y1.
Press **[MENU] 5** (GRAPH)
To clear any functions,
press **[F2:DEL][F1:YES]** to
clear an entry, then enter
[x,θ,T][x²][+] **3 [x,θ,T]**[−] **1 [EXE]**
[MENU] 1 (RUN)
Store the number 2 in memory x ,
2 [→]**[x,θ,T][SHIFT]**[↵]
Press **[VARS][F4:GRPH][F1:Y] 1 [EXE]**
the result is 9.

All expressions containing an x will be evaluated with $x = 2$ until this number stored in the x box is replaced with another number.

8. **Angles:**

Remember: to delete a screen menu, press **[EXIT]**
The type of measure for angles is set in the SET UP menu.
From MENU 1(RUN)
Press **[SHIFT][SET UP]** Use [▽] to highlight Angle.
The choices are in the Screen menu.
Press **[F1:Deg]** or **[F2:Rad]** then press **[EXE]**
After the MODE is set, use the operations in **[OPTN][F6:▷][F5:ANGL]** to change the units.

To enter an angle in Degrees while in Radian MODE:

ex. Press **[SIN] 30 [F5:ANGL][F1:∘][EXE]** .5

To enter an angle in Radians while in Degree MODE:

ex. Press **[SIN]** π **[aᵇ/c] 6 [2nd][F5:ANGL][F2:*r*][EXE]** .5

Use the angle in Degrees for the following examples.
From the RUN Menu, press **[OPTS][F6:▷]**
To change Rectangular coordinates to Polar coordinates:

ex. Press **[F5**:**ANGL][F6**:▷]**[F1**:**Pol(]** **1**,**1** [)]**[EXE]**

In Polar form $r \approx 1.4142...$, $\theta = 45°$

To change Polar coordinates to Rectangular coordinates:

Press **[F5**:**ANGL][F6**:▷]**[F2**:**Rec(]** $\sqrt{2}$, **45** [)]**[EXE]**

In Rectangular form $x = 1$, $y = 1$

If the MODE is Radians, enter the angle in Radians

Building Tables:

The Casio has a built in TABLE function used to evaluate functions for different values of x.

From the MAIN MENU choose **[MENU] 7** (TABLE)

The algebraic expression to be evaluated is stored in any Y = menu.

Press **[F5**:**RANG]** Set Start : beginning entry

End: last entry, and pitch.

The pitch is the step size between entries.

ex. Enter $x^2 + 3x - 1$ in Y1 in the Y = Menu.

Press **[F5**:**RANG]** set the following values:

[(–)] **4 [EXE]** for Start, **10 [EXE]** for End

0.1 [EXE] for pitch.

Press **[EXIT]** when finished.

Press **[F6**:**TABL]** to view the table.

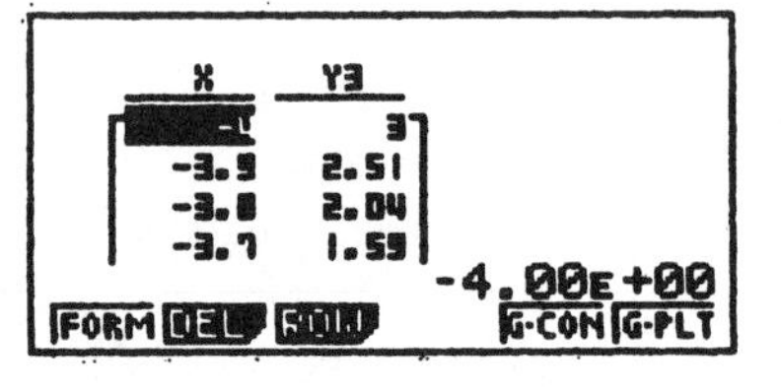

Use the up and down arrows to see more values in the table.

Tables may also be constructed in the STAT menu to examine and evaluate statistical data .

10. Building Lists:

Numbers can be stored in Lists. Choose **[MENU] 4** (LISTS)

List entries may be entered one-by-one or as a batch input.

ex. Enter the numbers 1, 3, 5, 7 in list **List 1**

Press **[MENU] 4** (LIST)

[F4:**DEL-A][F1**:**YES]** to clear a highlighted list.

The first position in List 1 is highlighted.

Press **1 [EXE] 3 [EXE] 5 [EXE] 7 [EXE]**

or Delete the entries in List 1, Highlight **LIST 1**

[SHIFT][{] **1** , **3** , **5** , **7** **[SHIFT]**[}]**[EXE]**

The numbers are entered in List 1.

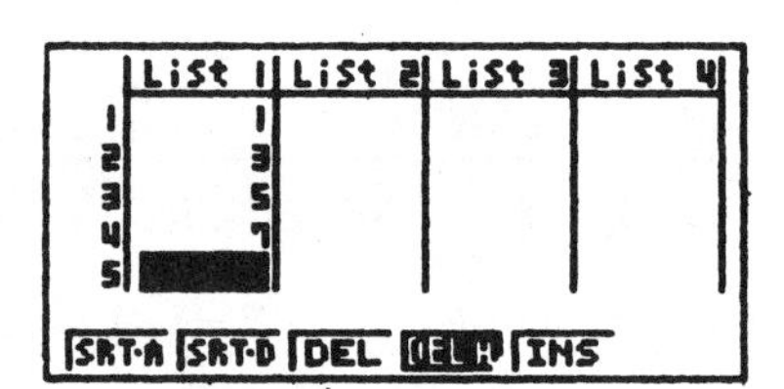

To display List1 on the home screen, press

[MENU] 1 (RUN)**[OPTN][F1**:**LIST]**

[F1:**List] 1 [EXE]**

The list is displayed as a Table.

A List can be entered onto the Home Screen
and stored as List 1
[SHIFT][{] 1 , 3 , 5 , 7 [SHIFT][}][→]
[OPTN][F1:LIST][F1:List] 1 [EXE] Done
The list is stored in MENU TABLE

```
{1,3,5,7}→List 1
                Done
```

Arithmetic operations are done using the +, −, ×, ÷, ^ keys,
when the lists are the correct lengths.
ex. To multiply a list by a constant:
[OPTN][F1:LIST] 3 [F1:List] 1 [EXE]
Other operations are done from the
RUN **[2nd][LIST][OPTN][F1:LIST]**.menus.

Special Lists:
Start from **[MENU] 1 [OPTN][F1:LIST]**
[F5:Seq(] expression, variable,begin, end, increment)
Generate a sequence of numbers using a rule for expression.
ex. **[F1:LIST][F5:Seq][(][x,θ,T][x²] , [x,θ,T],1, 5, 2]**
square the numbers x, starting with 1, adding 2, ending at 5.
The result is { 1, 9, 25}
Compute the Sum of the numbers in a sequence
ex. **[F1:LIST][F6:▷][F6:▷]F1:Sum][F6:▷][F1:List] 1 [EXE]**
The result is 16

Generate a list of Cumulative Sums for a List.
ex. **[F1:LIST][F6:▷][F6:▷][F3:Cuml][F6:▷][F1:List] 1 [EXE]**

For the list {1 3 5 7} = List 1 The display is $\begin{bmatrix} 1 \\ 4 \\ 9 \\ 16 \end{bmatrix}$

[7:∆List(] Generate a list of the differences of consecutive terms.
ex. **[F1:LIST][F6:▷][F6:▷][F5:∆] 1**

For the list {1 3 5 7} = List 1 The display is $\begin{bmatrix} 2 \\ 2 \\ 2 \end{bmatrix}$

This list is 1 entry shorter than the original list.

Summing Sequences:
The Σ command can be used to find the Sums of sequences.
Press **[MENU] 1 [OPTN][CALC][F6:▷][F3:**Σ(]
The command is Σ(general term, summation variable,
beginning term #, ending term #, partition size)

ex. $\Sigma_{k=2}^{6}(k^2 - 3k + 5)$
Press **[CALC][F3**:Σ(]$k^2 - 3k + 5, k, 2, 6, 1)$
[EXE] the answer is 55

11. Graphing Functions:

The function to be graphed must be entered
as one of the 20 functions in Main Menu **[MENU] 5** (GRAPH)
In the Y = menu, enter $x^2 + 3x - 1$ in Y1.

The function must be turned on to graph, so,
the = must be highlighted.
To turn the function on or off,
Highlight the function. Press **[F1:SEL]**
Each **[F1:SEL]** reverses the current state.
Color: There are 3 colors for graphs.
After entering the function in the Y = menu,
Highlight it. Press **[F4:COLR]**
Pick the color, the highlighted function
will be graphed in that color.
To graph the function press **[F6:DRAW]**
The function is graphed on the current window settings.

A graph may be translated up, down, right or left,by
pressing an arrow key before pressing any other keys.

The ZOOM menu has choices to zoom in or out
and for built in window sizes.

Press **[SHIFT][V-Window][F1:INIT][EXIT][F6:DRAW]**
This initializes the Decimal Window
The window is set to a Friendly window.
Each pixel is 0.1 units, and the graph is its
true shape. Unit size on X axis = unit size on Y axis.
Any multiple or translation of these settings
also makes a Friendly window.

Press **[SHIFT][V-Window][F3:STD][EXIT][F6:DRAW]**
This is the Standard window, it sets
X to [-10,10] and Y to [-10,10]. The units on the X and Y axis
are **not** the same size, so graphs are not their true shape.

Press **[SHIFT][F2:Zoom][F6:▷][F2:SQR]** Now the graph is its true shape again.
The window settings for the Y axis are left the same, and the X settings
have been recalculated so that the units on the X and Y axis are the
same. This is called a Square window. It usually is not Friendly.

Press **[SHIFT][F2:Zoom][F5:AUTO]** This is the AutoScale function.

After the Xmax, Xmin are set, it finds a Y-range to show a Complete Graph. The Y settings often need adjustment to get a clear picture. With AUTO a Friendly Window stays Friendly.

Press **[SHIFT][V-Window]** It displays the current settings for the X and Y ranges. These values can be changed by using the arrow keys to highlight the entry, then typing the new value.
Use **[EXE]** after each change.
Use the [▽] to pass by an unchanged entry.
(Remember to use (–) for a negative number).
The current values can be multiplied by a constant, or a constant can be added to an entry and the calculator will do the arithmetic.
To edit an entry in the **Y**= Menu press [▷] edit the entry, press **[EXE]**
Press **[EXIT][F6:DRAW]** to return to the graph.

Press **[SHIFT][F1:TRACE]** An orange blinking star cursor appears on the graph at the far left of the screen.
It may not be visible. It may be off of the screen..
The left and right arrow keys move this cursor along the function graph.
The coordinates that are displayed are points that satisfy the function equation. The Y values are computed from the X value of the pixel.
When more than one function is plotted, the up and down arrows move the cursor vertically between the different plots. These coordinates are only approximations unless the graph is on a Friendly window and the coordinate is a rational number.

To graph a <u>*split function*</u> $f(x) = \left\{ \begin{array}{ll} x+7, & x \le -5 \\ 4-x, & x > -5 \end{array} \right\}$

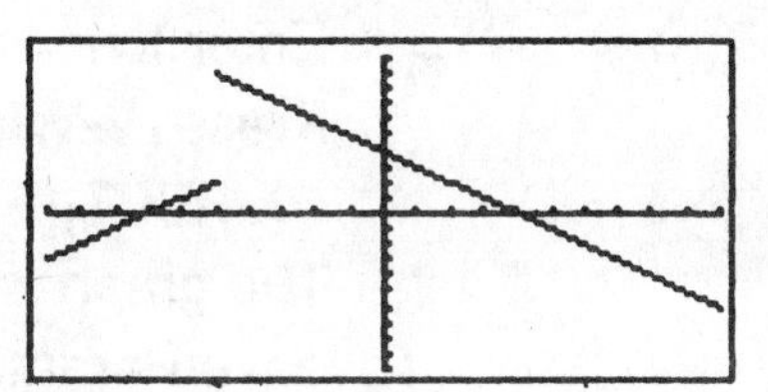

Press **[MENU] 5** (GRAPH). In the Y = menu enter the functions in 2 empty spaces.
$Y1 = (x+7)(x \le -5) \qquad Y2 = (4-x)(x > -5)$
$Y1$ =**[x,θ,T][+] 7** , [–**20**,–**5**]
$Y2$ = **4** [–]**[x,θ,T]** , [–**5**,**20**]
the -20, 20 are arbitrary, use any point off the graph.
Graph using **[SHIFT][V-Window][F3:STD][EXIT][F6:DRAW]**

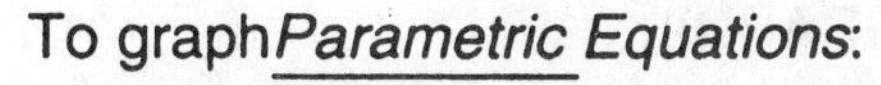
To graph *Parametric Equations*:

Parametric Equations are graphed in **[MENU] 5** (GRAPH).
From the Y = menu, press **[F3:TYPE]**
Different types of functions can be graphed on the same screen.
Press **[F3:Parm]** all of the empty spaces in the Y = menu are now set to Parametric mode.
The menu now reads $\mathbf{X_{1T}}$= and $\mathbf{Y_{1T}}$=
Now X and Y are both functions of

the independent variable T. Use the down arrow to move down to the T,θ View Window

The settings for min and max determine how much of the graph is plotted, pitch determines how many points are plotted. **[EXIT]**

Enter the functions and press **[F6:DRAW]**

ex. Write the function $y = x^2 + 3x - 1$ in parametric form.

Graph Func :Param
Xt1■T
Yt1■T²+3T−1
Yt2:
Xt3:
Yt3:
Y= r= Parm X=c

Let $x = t, \quad y = t^2 + 3t - 1.$

Enter $\mathbf{X}_{1T} = T$

$\mathbf{Y}_{1T} = T^2 + 3T - 1$

Press **[SHIFT][V-Window][F3:STD]**

This is the Standard window.

Go down to the T,θ View Window.

Set **min** = **-5**, **max** = **5**, **pitch** = **0.5**

[EXIT][F6:DRAW] The graph is displayed.

Use **[SHIFT][TRACE]** A point is traced every .5 unit

Now set the pitch to .1,

The x and T variables now trace at .1 unit per pixel.

This is a Friendly Window.

To graph *implicit functions* easily:

Write the equation in Parametric form.

Change GRAPH Type to Parametric.

ex. Write the equation $x^2 + y^2 = 4$ in parametric form: $x = 2\cos(t), y = 2\sin(t)$.

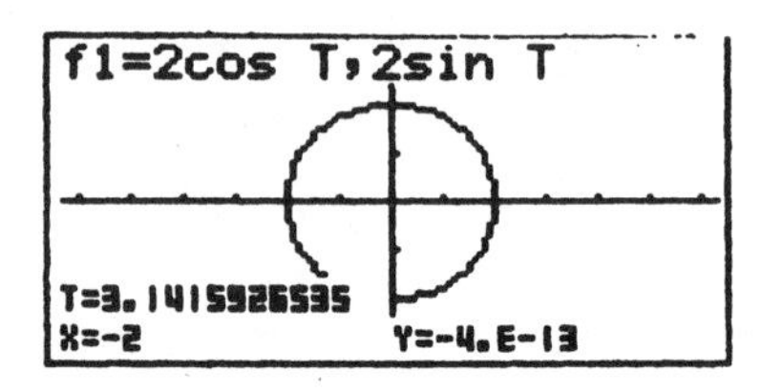

Enter $\mathbf{X}_{1T} = \mathbf{2}\cos(\mathbf{T}), \mathbf{Y}_{1T} = \mathbf{2}\sin(\mathbf{T})$

Press **[V-Window][F1:INIT]**

[EXIT][F6:DRAW]

The graph is a circle.

To leave Parametric Mode, **[EXIT][F3:TYPE]**

Change the GRAPH type with the screen menu keys.

To graph *Polar Equations*:

Polar Equations are graphed in **[MENU] 5** (GRAPH)

From the Y = menu, press **[F3:TYPE]**

Press **[F2:**r =] all of the empty spaces in the Y = menu are now set to Polar mode.

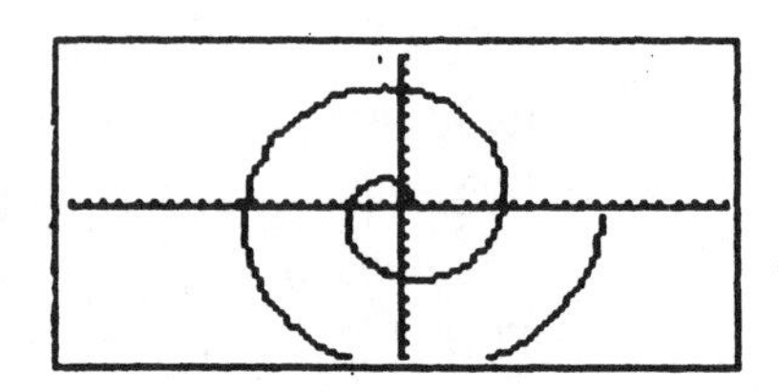

In the Y = menu, enter **r1** = θ, where r = f(θ).

Press**[SHIFT][V-Window][F3:STD]]**

Polar is a special case of Parametric,

Set $\theta\text{min} = 0$, $\theta\text{max} = 4\pi$, $pitch = \frac{\pi}{24}$

[EXIT][F6:DRAW]
[SHIFT][Zoom][F6:▷][F2:SQR]

To leave Polar Mode, **[EXIT][F3:TYPE]**
Change the graph type with the screen menu keys.

Graphing the Conics:

Press **[MENU] 9** (CONICS)
There are 9 common conic equations built into a menu.
Use the down arrow to highlight the equation
press **[EXE]** Enter the values of your constants on the line that is highlighted. **[EXE]** after each entry.
Pick a viewing window to be a Friendly Window that shows a complete graph. **[EXE]**
Remember the arrow keys translate the graph.
Press **[SHIFT][Zoom][F6:▷][F2:SQR]** The graph is now its true shape.

12. Solving Equations:

Write the equation in the form $f(x) = 0$
Using Trace and Zoom will give you an approximate answer.
Use either a Zoom box or use Zoom In.

To Set the ZOOM Factors:

After a graph is drawn,
Press **[F2:Zoom][F2:FACT]** and set the factors.
Press **4 [EXE] 4 [EXE][EXIT]**
(The default [INIT] settings are 2)

ex. Find the real zero of $x^3 + x + 1 = 0$
correct to hundredths. (Use both Zoom factors = 4)

Use Zoom In:

Press **[MENU] 5** (GRAPH) In the Y= menu
Enter **[x,θ,T][^] 3 [+][x,θ,T][+] 1 [EXE]**
[SHIFT][V-Window][F1:INIT]
[EXIT][F6:DRAW]

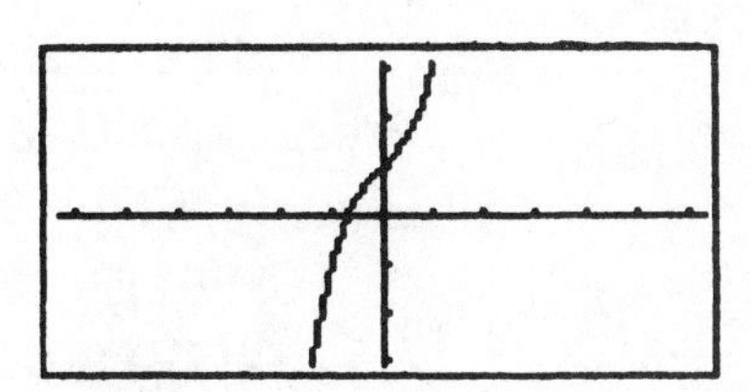

The graph crosses the X axis between -1 and 0.
Press **[V-Window]**
Set the Xscale = 0.01 **[EXE]**
[EXIT][F6:DRAW]
Press **[SHIFT][F1:TRACE]**
Move the cursor near the root.

[SHIFT][F2:Zoom][F3:IN] Repeat
[SHIFT][F1:TRACE]Cursor near the root
[SHIFT][F2:Zoom][F3:IN] Repeat
[SHIFT][F1:TRACE]Cursor near the root
[SHIFT][F2:Zoom][F3:IN] Repeat

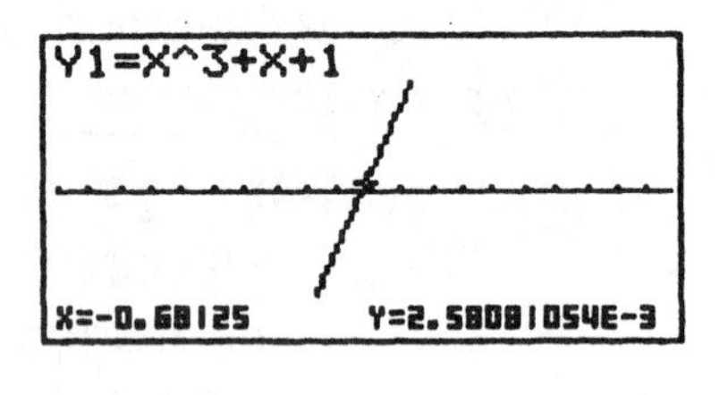

Now the tick marks are visible on the X-axis.
[SHIFT][F1:TRACE]
For $x \approx -.6828125, y \approx -.0011622$
For $x \approx -.68125, y \approx .00258081.$
The value of the zero is $\underline{x \approx -.68}$ correct to hundredths..

Note that these 2 values of y lay on opposite sides of the X axis.
(These figures depend on the value of the Zoom Factors, these are set at 4).

Use Zoom Box:

Press **[MENU] 5** (GRAPH) In the Y= menu
Enter **[x,θ,T][^] 3** [+]**[x,θ,T]**[+] **1 [EXE]**
[SHIFT][V-Window][F1:INIT]
[EXIT][F6:DRAW]
You can see that the function crosses the X axis between -1 and 0.
Press **[V-Window]**
Set the Xscale = 0.01 **[EXE]**
[EXIT][F6:DRAW]
[SHIFT][F2:Zoom][F1:BOX]
Move the cursor with the arrow keys until it is above and to the left of the zero. Press **[EXE]**
Use [▷], then [▽] to draw a box with the zero inside **[EXE]**
Repeat until the tick marks on the X axis are clearly visible.
Trace to approximate the value $x \approx -.68...$

Using the Graph Solve feature:

ex. Find the real zero of $x^3 + x + 1 = 0$ correct to hundredths.
Press **[MENU] 5** (GRAPH) In the Y= menu
Enter **[x,θ,T][^] 3** [+]**[x,θ,T]**[+] **1 [EXE]**
[SHIFT][V-Window][F1:INIT]
[EXIT][F6:DRAW]

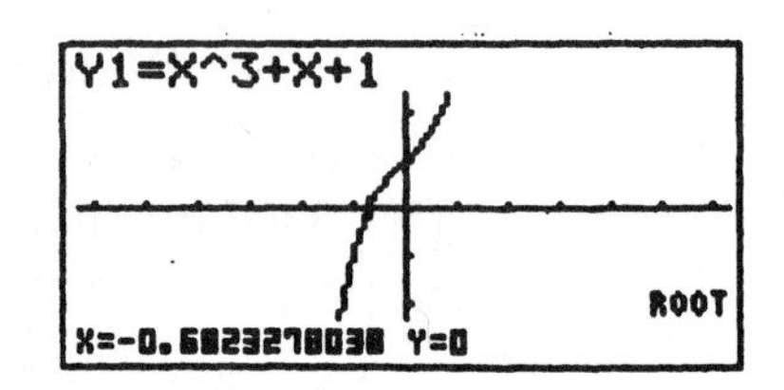

The graph crosses the X axis between -1 and 0.
Press **[SHIFT][G-Solv][F1:ROOT]**
The graph is drawn. ***Wait***
The orange cursor moves up the graph and

Stops on the intersection.
The approximation $x \approx -.682...$ correct to 3 places
is displayed on the bottom of the screen.

Using the Numeric Solver.

ex. Find the real zero of $x^3 + x + 1 = 0$ correct to hundredths.
Press **[MENU] 5** (GRAPH) In the Y= menu
Enter **[x,θ,T]**[^] **3** [+]**[x,θ,T]**[+] **1 [EXE]**
[SHIFT][V-Window][F1:INIT]
[EXIT][F6:DRAW]
The graph crosses the X axis
between -1 and 0.
Press **[MENU] 1** (RUN) **[OPTN]**
[F4:CALC][F1:Solve]
Enter (the function, guess, lower bound, upper bound)
[x,θ,T][^] **3** [+]**[x,θ,T]**[+] **1** , **-.5** , **-1** , **0** **[)][EXE]**
The x value is now $x \approx -.682...$ correct to hundredths.
(A guess of -1 or 0 would have worked as well. Try it.)

Using the Polynomial Solver.

If the problem is to find the zeros of a 2nd or 3rd degree polynomial,
use the POLYNOMIAL SOLVER
ex. Find all of the zeros of $x^3 + x + 1 = 0$
Press **[MENU] A** (EQUA)
Press **[F2:POLY][F2:3]**
Fill in the grid, **1 [EXE] 0 [EXE] 1 [EXE] 1 [EXE]**

[F1:SOLV] the ans. is displayed $\begin{bmatrix} 0.3411 + 1.1615i \\ 0.3411 - 1.1615i \\ -0.682 \end{bmatrix}$

Using the Equation Solver.

ex. Find the solutions to $\tan(x) = 0$.
Press **[MENU] A** (EQUA) **[F3:SOLV]**
After Eq: enter **[tan][X,θ,T][SHIFT][=] 0 [EXE]**
After X = enter a guess
(do this even if the line is highlighted)
2 [EXE][F6:SOLV]
The value of X changes to $X = 3.1415... \approx \pi$

Solving Simultaneous Linear Equations:

Press **[MENU] A** (EQUA) **[F1:Simultaneous]**

Press **[F2:3]** Enter the coefficients.
Press **[F1:SOLV]**

ex. Solve the set of equations:
$x + y + 2z = 1$
$-x + y = 1$
$2x + y + z = 0$
Enter the coefficients in the matrix:

$$\begin{bmatrix} 1 & 1 & 2 & 1 \\ -1 & 1 & 0 & 1 \\ 2 & 1 & 1 & 0 \end{bmatrix} \quad \textbf{[F1:SOLV]} \quad \begin{matrix} x = \frac{-1}{2} \\ y = \frac{1}{2} \\ z = \frac{1}{2} \end{matrix}$$

Finding the Intersection of 2 graphs:

ex. Find the intersections of $y = x^2 + 2x - 3$ and $y = \frac{x}{2}$.
Press **[MENU] 5** (GRAPH) In the Y= menu
Enter $Y1$ = **[x,θ,T][^] 3** [+]**[x,θ,T]**[+] **1 [EXE]**
$Y2$ = **[x,θ,T]**[$a^b/_c$]**2 [EXE][SHIFT][V-Window]**
[F1:INIT][EXIT][F6:DRAW]
You can see the intersections are in the first and third quadrants. Press **[SHIFT][G-Solv]** **[F5:ISCT]** There is a box on the upper right end of the line. Press **[EXE]** ***Wait.***
The orange blinking cursor moves up to the intersection. The display on the screen is
$x \approx -2.637...$ $y \approx -1.318...$
Now Press [▷] ***Wait*** The orange blinking cursor moves to the other intersection.
The display on the screen is
$x \approx 1.137...$ $y \approx 0.569...$

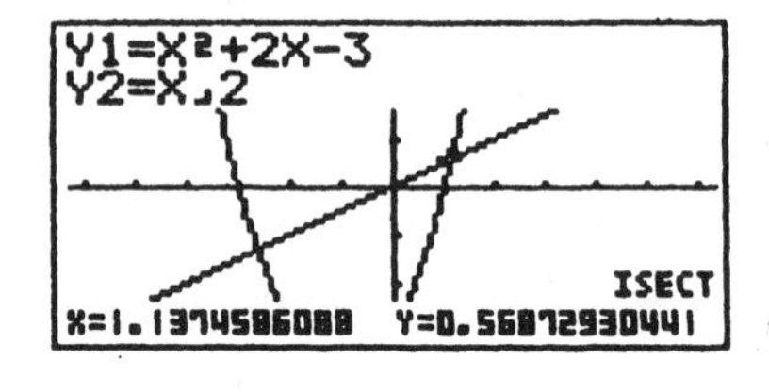

13. Finding Maximum and Minimum values of a Function:

The X value is the location of the extrema.
The Y value is the value of the extrema.
Find the minimum value of $f(x) = x^2 + 3x - 1$.

Using Graph Solve:

Press **[MENU] 5** (GRAPH). In the Y= menu
Enter $Y1$ = **[x,θ,T]**[x^2][+] **3 [x,θ,T]**[–] **1 [EXE]**
The minimum is off the screen.
Press [▽] The function is regraphed.
Press **[SHIFT][G-SOLV][F3:MIN]** ***Wait***
The orange blinking cursor moves to the minimum

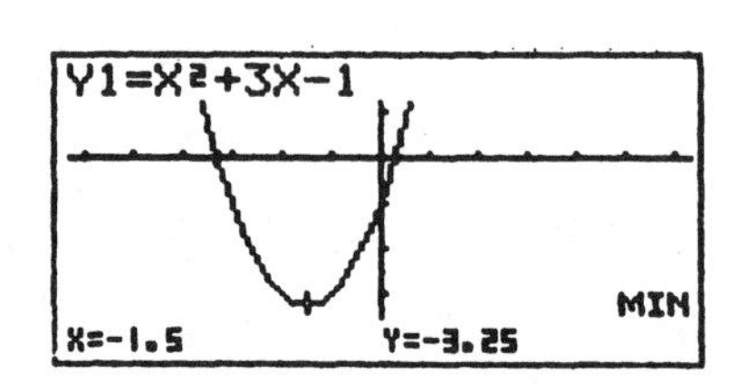

The display reads $x = -1.5 \quad y = -3.25$.

Using the Solver:

This algorithm is iterative and requires a lower and upper x bound for the minimum. These can be estimated from the graph.

For $f(x) = x^2 + 3x - 1$, we can use a lower bound of $x = -2$, and an upper bound of $x = -1$. Press
[MENU] 1 (RUN) **[OPTN][F4:CALC][F6:▷][F1:FMin]**
put (expression, lower bound, upper bound)

The screen shows **Fmin**$(x^2 + 3x - 1, -2, -1)$ **[EXE]**

The answer displayed is $\begin{bmatrix} -1.5 \\ -3.25 \end{bmatrix}$ where the first number is $x = -1.5$, the second number is $y = -3.25$.

14. Statistic Plots:

Entering Data:

Press **[MENU] 2** (STAT) to display the list editor.
Clear all data from each list by highlighting any entry in the list, then pressing **[F3:DEL-A][F1:YES]**
Use the following data:

x	1	2	3	4	5	6	7	8
y	1.1	2.6	3.8	5.1	5.9	7.2	8.2	9

Enter the independent variables, x, in list *List*1.
Highlight the first position in *List*1.
Press **1 [EXE] 2 [EXE]**...
Enter the dependent variables, y, in list *List*2.
Highlight the first position in *List*2.
Press **1.1 [EXE] 2.6 [EXE]**...

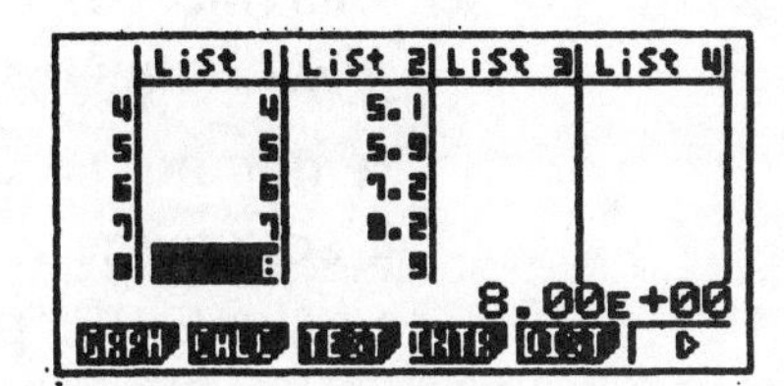

Constructing a Graph:

Press **[F1:GRPH][F4:SEL]**
Highlight StatGraph1 press **[F1:ON][F6:DRAW]**
A Scatter Plot is displayed, the regression choices are in the screen menu.
(Note:) The Scatter Plot using *List*1 and *List*2 is the default plot. See your manual to change the settings for other plots.

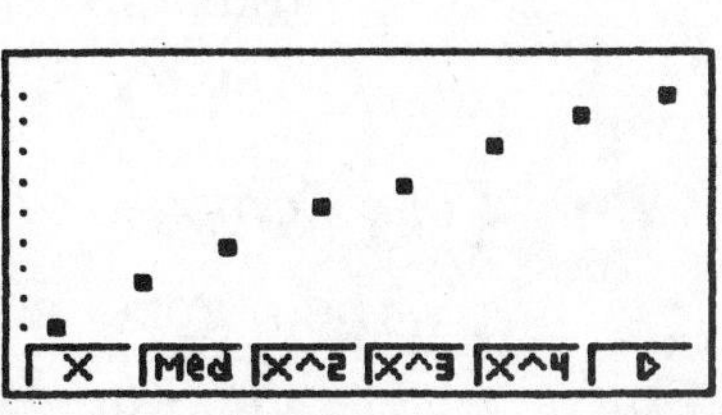

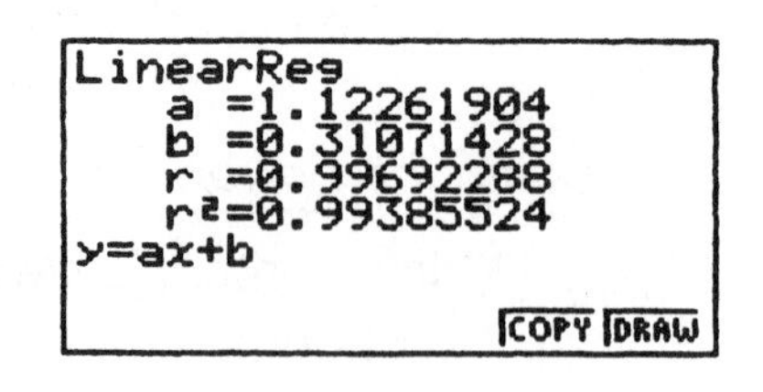

Regression Analysis:

From the Statistics Plot screen

Press **[F1**:×] to display
the Linear Regression

The regression coefficients and
equation are displayed.

Regression Plot:

A regression analysis is needed
before a Plot can be drawn.

From the Regression Analysis screen,
press **[F5:COPY]**

The Y= screen is displayed.

Move to the space where the
regression equation
is to be pasted. **[EXE]**

The current regression equation is
copied to the Y = menu.

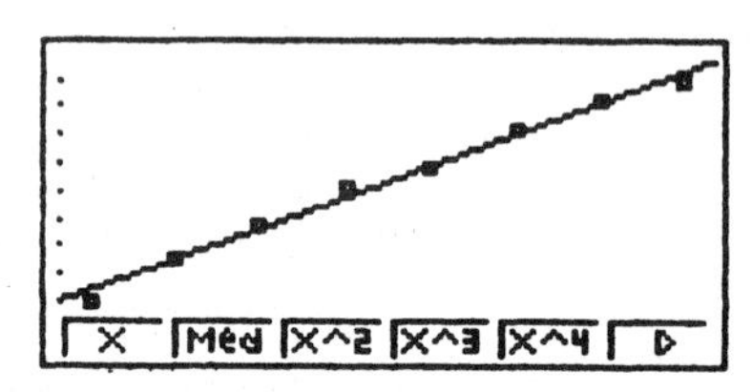

The Regression Analysis
screen is displayed.

Press **[F6:DRAW]**

Watch the regression line plot through the scatter plot.

15. Complex Numbers:

Rectangular Complex:

Press **[MENU] 1** (RUN) **[OPTN][F3:CPLX]**
to display the Complex Number Menu.

The number $i = \sqrt{-1}$ is **[F1;i]** located on the
screen menu.

To enter 2 + 3i, press **2** [+] **3 [F1:i]**

To multiply or divide complex numbers, each number
must be entered in a parenthesis.

ex. (**2** + **3i**) + (**3** – **5i**) = **5** – **2i**

Polar Complex:

The number a + bi is written $re^{i\theta}$

$r = \sqrt{a^2 + b^2} \quad \theta = \arctan(\frac{y}{x})$

To change a complex number to polar form:

Press **[EXIT][F6:▷][F5:ANGL][F6:▷]**
[F1:**Pol(] 3 , 4 [EXE]**

The display reads $\begin{bmatrix} 5 \\ .9272 \end{bmatrix}$ where $r = 5$

and the angle $\theta = .9272\ldots radians$

Complex Functions:

The list of functions and operations for complex numbers is displayed in the **[OPTN][CMPLX]** screen menu.

16. Matrices:

Matrices are *stored* by name, edited and used in matrix arithmetic in **[MENU] 3** (MAT)

The menu holds 26 matrices named A - Z and an Ans. matrix to hold the result of a calculation.

To enter a matrix, highlight the matrix name, press [▷]

Enter the dimensions, rows **[EXE]** columns **[EXE]** then fill in the matrix using **[EXE]** after each entry.

ex. Enter a 2 row, 2 column matrix as MAT A

Highlight Mat A, press [▷] **2 [EXE] 2 [EXE]**

A 2 × 2 grid is displayed with element a_{11} highlighted.

1 [EXE] 2 [EXE] 3 [EXE] 4 [EXE]

The matrix is saved as mat A = $\begin{bmatrix} 1 & 2 \\ 3 & 4 \end{bmatrix}$

Matrix *operations* are displayed in **MENU 1** (RUN)

From this screen press **[OPTN][F2:MAT]**

Matrix *arithmetic* +, –, ×, ^ is done with the arithmetic keys on the keyboard.

Matrix *Row Opreations* are displayed with **[F1:R-OP]** on the Matrix Entry screen.

This Casio does handle fraction arithmetic while row reducing a matrix.

See the 9850 manual for details.

Matrices may be *entered from the keyboard in MENU* 1:

Enter and store: $\begin{bmatrix} 1 & 1 & 2 & 1 \\ -1 & 1 & 0 & 1 \\ 2 & 1 & 1 & 0 \\ 1 & 3 & 1 & 0 \end{bmatrix}$ as matrix A.

Use [(–)] for a negative number.

[SHIFT][[ ][SHIFT][[ ]1 , 1 , 2 , 1 [SHIFT][ ]]

[SHIFT][[ ] -1 , 1 , 0 , 1[SHIFT][ ]]
[SHIFT][[ ] 2 , 1 , 1 , 0 [SHIFT][ ]]
[SHIFT][[ ] 1 , 3 , 1 , 0 [SHIFT][ ]]
[SHIFT][ ]][→][OPTN][F2:MAT]
[F1:Mat][ALPHA] A [EXE]

To find the inverse:

[OPTN][F2:MAT][F1:Mat][ALPHA] A
[SHIFT][x^{-1}][EXE]
[→][OPTN][F2:MAT]
[F1:Mat][ALPHA] B [EXE]

To check the inverse:

[F2:MAT][F1:Mat][ALPHA] A [×]
[F2:MAT][F1:Mat][ALPHA] B [EXE]
The answer is the Identity of order 4.

To find the determinent of matrix A:

[F2:MAT][F3:Det][F1:Mat] 1 [EXE]
The answer is 6.

Notes:

This Casio does not find row echelon and reduced row echelon forms unless they are done with row operations.
Programs for rref are available on the CASIO web site.
Systems of equations with unique solutions (up to 6 variables) can be solved in the EQUATION MENU.
If any entry in the matrix is a rational number, the reduced form will have decimals in the matrix, but the entry displayed at the bottom of the screen will be a fraction.

Chapter 8 – Using the HP-38G Graphing Calculator.

1. Notation:

Keystrokes, except for numbers and commas are in bold brackets **[MATH]**

For a second function x^2, press **[2nd]** [x^2] the second functions are printed in blue over the keys.

For the letter A, press **[A...Z][A]** the ALPHA symbols are printed in tan on the right under the keys.

The top-row keys are called menu keys beecause their meanings depend on the context, that is why they are blank. The bottom line of the display shows the labels for the menu key's current meanings. Press {{**STO▸**}} means to press the blank menu key under the screen label STO▸.

Calculator results are underlined.

2. Basics:

To adjust the screen contrast: Simultaneously press **[ON]** and [+] to increase the contrast or **[ON]** and [–] to decrease the contrast.

Mode settings: Press ■**[MODE]** select Radians, Standard, Dot. To change a mode press {{**CHOOS**}} highlight the new setting using the down arrow, then press {{**OK**}} or use [+] to cycle through the choices.

Other important settings are in the MATH and LIB menus.

HOME displays the home screen where you calculate. There are 4 lines to show history: the most recent input and output and an edit line at the bottom. Data is input on the edit line. After you press ENTER, it is displayed as history. When you EDIT, it returns to the edit line.

An APLET is a built in application to solve certain types of problems. There are 6 built in aplets stored in LIB, more are available from the HP web site.

3. Arithmetic Computations:

Order of Operations: The order is Algebraic, operations are performed from left to right. First exponentiation, then multiplication and division, then addition and subtraction. Parenthesis must be used to change the algebraic order. Parenthesis must be used around the numerator and denominator in Algebraic Fractions.

Minus sign: Use [-x] for negative, [–] for subtraction.

Fractions: to enter fractions use [÷]. If the NUMBER FORMAT in the MODE menu is set for Fraction, the quotients will be displayed as fractions.The last answer can be changed by changing to

To calculate: press [**ENTER**]

ex. $2 \times 3 + 4 \times 5$ [**ENTER**] 26
$2 \times (3 + 4) \times 5$ [**ENTER**] 70
$3 + 4 \div 2$ [**ENTER**] 5
$(3 + 4) \div 2$ [**ENTER**] 3.5
Change the MODE to Fraction. Return to the home screen.[**HOME**]
[△] Highlight 3.5 {{**COPY**}}[**ENTER**] The display reads 7/2

4. **Scientific Notation**:

From the keyboard use number between 1 and 10 ■[**10^x**] exponent.
The exponent must be a number between -99 and 99.
To Display answers in Scientific Notation change the MODE menu entry.
■[**MODE**]{{**CHOOS**}}{{**OK**}} Press [▸] to change the number of decimal places in the display.
To return to Normal:
■[**MODE**] Highlight Scientific {{**CHOOS**}} highlight Standard, press {{**OK**}} [**HOME**]

5. **Editing**:

Before typing [***ENTER***]

A blinking arrow determines the current position on the screen.
The arrow keys move the cursor around the screen.
Edit is in Insert mode.

To delete a symbol, move to the symbol, press [**DEL**]
To change an entry move the blinking cursor to the entry, press [**DEL**] and type the new entry.
To insert a symbol, move to the symbol after the insertion point, then type the new text.

After typing [***ENTER***]

Highlight the command in history, press {{**COPY**}} the command is returned to the edit line. .Edit it as explained above in *Before typing ENTER*
To repeat the last line, press [**ENTER**]
To clear the edit line, press [**ON**] (CANCEL)
To clear the home screen press ■[**CLEAR**]
To exit a MENU, press {{**CANCL**}} or [**HOME**]
To erase a function from the Y = screen, highlight the function, press [**CLEAR**]

6. **Entering Algebraic Functions:**

Algebraic functions must be entered into the application screen you need to use. After you pick an application, press **[SYMB]** the proper screen will be displayed.
From the aplet menu in **[LIB]** highlight **Function [ENTER]**
Use the arrow keys to move to an empty position.
Enter the function. It is now available to use as a graph or a table. This menu holds 10 functions.When Function mode is set, access
this screen by pressing **[SYMB]**

On the home screen, enter the function on the edit line. When you press **[ENTER]** it will be evaluated with the values of the variables that are stored in Memory.

Press ■**[CHARS]** this is a list of all symbols in the calculator.
Highlight the symbol, press {{**OK**}} and the symbol is pasted onto the home screen.

When you enter built in functions, left parenthesis are included. Complete the input variable, then type the right parenthesis.

The key **[x,T,θ]** prints x in Function mode, T in parametric mode and θ in polar mode.

Special Functions:

Absolute value of x: press ■**[ABS][x,T,θ]**

Powers, x^5: press **[x,T,θ][x^y] 5**

2^x: press **2 [x^y][x,T,θ]**

e^x: press ■**[e^x][x,T,θ]** the ^ is forced.

Roots, $\sqrt[5]{x}$: press **[x,T,θ][x^y][(]1 [÷] 5 [)]**

Natural log, $ln(x+5)$: press■ **[ln][x,T,θ][+] 5 [)]**

Conjugate Pairs, $\pm\sqrt{4-x^2}$: enter **F1(x)=[+][$\sqrt{\ }$][(] 4 [−] [x,T,θ][x^2][)]** and**F2(x)=[−][$\sqrt{\ }$][(] 4 [−] [x,T,θ][x^2][)]**

Factorial, 5!: press **5 [MATH] P [▼][▸]** highlight ! {{**OK**}}

Trig function, $tan(3x^2)$: press **[TAN] 3 [x,T,θ][x^2][)]**

Inverse trig function, $arcsine(x)$: press ■**[ASIN][x,T,θ][)]**

Combinations, ${}_5C_2$: press **[MATH] P [▼][▸] [COMB]{{OK}} 5 , 2 [ )]**

7. **Evaluating a function: (checking your answer)**

Variables are represented as single alphabetic characters, A through Z, and θ. Numbers are stored in these positions using the command {{**STO▸**}} an arrow ▸ is shown on the home screen.

When an algebraic expression is entered, the variables are replaced with the stored constants and a number is displayed.

To evaluate a function on the home screen use the key **[ALPHA]** [:]to connect commands.

ex. To evaluate $x^2 + 3x - 1$ for $x = 2$,
Press **2 [STO▸][x,T,θ,n][ALPHA][:]**
[x,T,θ,n][x²][+] 3 [x,T,θ,n][−] 1
[ENTER]
the result is 9.

If the function is stored in the Function menu it may be recalled from the VAR menu.

ex. Enter the function $x^2 + 3x - 1$ in the Function menu as F1(x).
Press **[SYMB]** press **[ON]** then
[x,T,θ][x²][+] 3 [x,T,θ][−] 1 [ENTER][HOME]
Store the number 2 in memory x **2 [STO▸][x,T,θ][ENTER]**
Press **[A...Z][F] 1 [( ][x,T,θ][ )][ENTER]**
the result is 9.

All expressions containing an x will be evaluated with $x = 2$ until this number stored in the x box is replaced with another number.

8. Angles:

The type of measure for angles is set in the MODE menu.
Set ■ **[MODES]** Angle measure to either **Radians** or **Degrees** by highlighting the entry and pressing {{**OK**}}
To change an angle in Degrees to Radians
Press **[MATH][Real][DEG→RAD] 30 [ENTER]**
To change an angle in Radians to Degrees
Press **[MATH][Real][RAD→DEG] π ÷ 6 [ENTER]**

9. Building Tables:

The HP38 has a built in TABLE function used to evaluate functions for different values of x. The algebraic expression to be evaluated is stored in the Function **[SYMB]** menu.

ex. Enter $x^2 + 3x - 1$ in **[SYMB]** F1(X)
In the **■[NUM][SET UP]**menu, set
NUMSTART: -4
NUMSTEP: 1
NUMTYPE: Automatic
NUMZOOM 1
Press **[NUM]**

Use up and down arrows to see more values in the table.
Tables may also be constructed in the STAT menu to

examine and evaluate statistical data.

10. **Building Lists**:

Numbers can be stored in Lists either on the home screen, the LIST CATALOG or in the STAT EDIT menu.

Lists may be named (stored) using the names

L1, L2, L3, L4, L5, L6, L7, L8, L9, L0

ex. Enter the numbers 1, 3, 5, 7 in list **L1**

■[{] 1 , 3 , 5 , 7 ■ [}][STO▸] L1[ENTER]

Lists from the keyboard are stored in the **■ [LIST]** List Catalog

To display L1 on the home screen, press **[A...Z] L 1 [ENTER]**

Arithmetic operations are done using the +,–,×,÷,^ keys, when the lists are the same length.

Other operations are done from the **[MATH]** Highlight **List** [▸] menu.

Special Lists:

[MAKELIST(] expression, variable, begin, end, increment)

Generate a sequence of numbers using a rule for expression.

ex. **[MATH]** Highlight **[List][▹] [MAKELIST]**

[x,T,θ][x²] , [x,T,θ] , 1 , 5 , 2][)][ENTER]

square the numbers starting with 1, adding 2, ending at 5.

The result is { 1, 9, 25}

ΣLIST5(] (list) returns the sum of all the numbers in the list. For L1, the sum is 16.

△List(] (list) returns a list of the differences of consecutive terms.

For L1, the list is { 2 , 2 , 2 }

This list is 1 entry shorter than the original list.

11. **Graphing Functions**:

The function to be graphed must be entered as one of the 10 functions

In the **[LIB]** highlight **Function [SYMB]**

Enter $x^2 + 3x - 1$ in F1(X).

The function must be turned on to graph, so,

press **{{√CHK}}**

To turn the function on or off,

highlight the function and press **{{√CHK}}**

Each **{{√CHK}}** reverses the current state.

To graph the function press key **[PLOT]**

The function is graphed on the current window settings.

Press **{{MENU}}{{ZOOM}}** for choices to zoom in or out.

Press **■[VIEWS]** to find built in window sizes.

Highlight **Decimal {{OK}}** The graph is redrawn on the Decimal window.

The window is set to a Friendly window
Each pixel is 0.1 units, and the graph is its true shape. Unit size on X axis = unit size on Y axis.

Any multiple or translation of these settings stays a Friendly window.

Press ■ **[PLOT]** SET UP It displays the current settings for the X and Y ranges. These values can be changed by using the arrow keys to highlight the entry, then typing the new value.
Use **[ENTER]** after each change.
This does not usually give a Friendly window.
Use **[PLOT]** to return to the graph.

Press {{**MENU**}}{{**ZOOM**}} Highlight **Square** {{**OK**}}
Now the graph is its true shape again.
The window settings for the Y axis are left the same, and the X settings have been recalculated so that the units on the X and Y axis are the same. This is called a Square window. It usually is not Friendly.

Press ■**[VIEWS]** Highlight **Auto Scale**
After the Xmax, Xmin are set, it finds a Y-range to show a Complete Graph. The Y settings often need adjustment to get a clear picture. With Auto Scale a Friendly Window stays Friendly.
(Remember to use (–) for a negative number).
Press **[PLOT]** to return to the graph.

Be sure that {{**TRAC□**}} contains the white box. If not, press the menu key. The white box will appear. When the box is in the screen menu, TRACE is turned ON.

Use the left and right arrow keys move this cursor along the function graph.

The coordinates that are displayed are points that satisfy the function equation. The Y values are computed from the X value of the pixel.

When more than one function is plotted, the up and down arrows move the cursor vertically between the different plots. These coordinates are only approximations unless the graph is on a Friendly window and the coordinate is a rational number.

To graph a *split function*: $f(x) = \left\{ \begin{array}{ll} x+7, & x \leq -5 \\ 4-x, & x > -5 \end{array} \right\}$

Press **[SYMB]** then enter the function in 2 empty spaces.
$Y1 = (x+7) \div (x \leq -5)$ $\quad Y2 = (4-x) \div (x > -5)$
$Y1$ = [(][**x,T,θ**][+] **7** [)][÷]
[(][**x,T,θ**]■**[CHARS]** Highlight ≤ [(-x)] **5** [)]
$Y2$ = [(] **4** [–][**x,T,θ**][)][÷][(][**x,T,θ**]
■**[CHARS]** Highlight > [(-x)] **5** [)]
Graph using [**–10, 10**] by [**–10, 10**]

To graph *Parametric Equations*: press **[LIB]** arrow down and

highlight Parametric, then press **[ENTER]**
Press **[SYMB]** the menu now reads **X1(T)** = and **Y1(T)** =
You may want to change the window settings.
Now X and Y are both functions of
the independent variable T.
The settings for TRNG determine how much of
the graph is plotted.
Tstep determines how many points are plotted.
Enter the functions in **[SYMB]**and press **[PLOT]**

ex. Write the function $y = x^2 + 3x - 1$
in parametric form.
Let $x = t, \quad y = t^2 + 3t - 1.$
Enter **X1(T) = T**
Y1(T) = T²+3T – 1
Press **■[PLOT]SET UP**
Set **TRNG [-x] 5 , 5**
Press **■[VIEWS] Decimal** The graph is displayed.
Use TRACE This is **not** a Friendly Window

To graph *implicit functions* easily:
Write the equation in Parametric form.
ex. Write the equation $x^2 + y^2 = 4$
in parametric form: $x = 2\cos(t), \quad y = 2\sin(t).$
Enter **X1(T)= 2cos(T), Y1(T) = 2sin(T)**
■[VIEWS] Highlight **Decimal**
The graph is a circle.
Remember to change back to Function Mode.

To graph *Polar Equations*: press **[LIB]**, arrow down and
Highlight Polar, then press **[ENTER]**
Press [**SYMB**, Enter **R1**(θ) = θ, where r = f(θ).
Polar is a special case of Parametric, so
in **■[PLOT]SET UP**
Set θRNG 0 and 4π
θSTEP $\pi \div 24$
XRNG -10 10
YRNG -10 10
[PLOT]
Remember to change back to Function Mode.

12. Solving Equations:

Write the equation in the form $f(x) = 0$
Using Trace and Zoom will give you an approximate answer.

Use either a Zoom box or use Zoom In.

To Set the ZOOM Factors:

Press **[ZOOM]**[▷]**[4:SetFactors]** and set the factors.

ex. Find the real zero of $x^3 + x + 1 = 0$
correct to hundredths. (Use both Zoom factors = 4)
[LIB] highlight **Function [ENTER]**

Use Zoom In:

In FUNCTION SYMBOLIC VIEW, **[SYMB]** in an empty space
Enter **[x,T,θ][x^y] 3** [+]**[x,T,θ]**[+] **1**
■**[VIEWS]** Highlight **Decimal** {{**OK**}}
The function crosses the X axis
between -1 and 0.
Press ■**[PLOT] SET UP**
Set the Xscl = 0.01 {{**OK**}} **[PLOT]**{{**MENU**}}
Check that TRACE is ON, [TRAC□]
Move the cursor near the root.
{{**ZOOM**}} Highlight **IN 4×4** {{**OK**}}
Move the cursor near the root.
{{**ZOOM**}} Highlight **IN 4×4** {{**OK**}}
Move the cursor near the root.
{{**ZOOM**}} Highlight **IN 4×4** {{**OK**}}
Now the tick marks are visible on the X-axis.
Move the cursor near the root.
{{(x,y)}}
Use the arrow keys
For $x \approx -.6828125,\ y \approx -.0011622$
For $x \approx -.68125,\ y \approx .00258081$.
The value of the zero is $\underline{x \approx -.68}$ correct to hundredths.
Note that these 2 values of y lay on opposite sides of the X axis.
(These figures depend on the value of the Zoom Factors, these are set at 4).

Use Zoom Box:

In FUNCTION SYMBOLIC VIEW, **[SYMB]** in an empty space
Enter **[x,T,θ][x^y] 3** [+]**[x,T,θ]**[+] **1**
■**[VIEWS]** Highlight **Decimal** {{**OK**}}
The function crosses the X axis
between -1 and 0.
Press ■**[PLOT] SET UP**
Set the Xscl = 0.01 {{**OK**}} **[PLOT]**{{**MENU**}}
Move the free cursor with the arrow keys until it is above and to the left of the zero. Press **[ZOOM]**

Highlight **BOX** {{**OK**}}
Move the cursor to a corner of the box. {{**OK**}}
Use [▷], then [▽] to draw a box with the zero inside.{{**OK**}}
Repeat until the tick marks on the X axis are clearly visible.
Trace to approximate the value $x \approx -.68$

Using the Graph Solve feature:

ex. Find the real zero of $x^3 + x + 1 = 0$ correct to hundreths.
Press [**SYMB**] In FUNCTION SYMBOLIC VIEW, in an empty space,
Enter [**x,T,θ**][**x^y**] **3** [+][**x,T,θ**][+] **1**
■[**VIEWS**] Highlight **Decimal** {{**OK**}} The function is graphed.
The function crosses the X axis
between -1 and 0.
Press [**2nd**][**CALC**][**2:zero**]
{{**MENU**}}{{**FCN**}} Highlight **Root** {{**OK**}}
The approximation is $x \approx -.682$, correct to 3 places.

Using the Numeric Solver:

ex. Find the real zero of $x^3 + x + 1 = 0$ correct to hundreths.
Press [**LIB**] Highlight [**SOLVE**][**ENTER**]
In FUNCTION SYMBOLIC VIEW,
Enter [**x,T,θ**][**x^y**] **3** [+][**x,T,θ**][+] **1**
■[**VIEWS**] Highlight **Decimal** {{**OK**}}
The function crosses the X axis
between -1 and 0.
Press [**LIB**] Highlight **Solve**
In SOLVE SYMBOLIC VIEW,
Enter [**x,T,θ**][**x^y**] **3** [+][**x,T,θ**][+] **1** {{=}} **0**
[**NUM**] Enter a guess **0** {{**OK**}}{{**SOLVE**}}
The x value is now $x \approx -.682...$ correct to hundreths.
A guess of -1 or -.5 would have worked as well. Try it.

Finding the Intersection of 2 graphs:

ex. Find the intersections of $y = x^2 + 2x - 3$ and $y = \frac{x}{2}$.
Press [**LIB**] highlight **Function**
$Y1$ = [**x,T,θ**][x^2][+] **2** [**x,T,θ,n**][−] **3**
$Y2$ = [**x,T,θ**][÷]**2** ■[**VIEWS**] Highlight **Decimal**
You can see the intersections are in the first
and third quadrants. TRACE to the intersection
in Quad. 1. Using [△] and[▽] arrows you can see the
coordinates are not "nice" numbers, so use an
INTERSECTION SOLVER.
Press {{**MENU**}}TRACE to the
intersection in Quadrant 1.

The function crosses the X axis
between -1 and 0.
Press ■**[PLOT] SET UP**
Set the Xscl = 0.01 **{{OK}} [PLOT]{{MENU}}**
Move the free cursor with the arrow keys until it is above and to the left of the zero. Press **[ZOOM]**
Highlight **BOX {{OK}}**
Move the cursor to a corner of the box. **{{OK}}**
Use [▷], then [▽] to draw a box with the zero inside.**{{OK}}**
Repeat until the tick marks on the X axis are clearly visible.
Trace to approximate the value $x \approx -.68$

Using the Graph Solve feature:

ex. Find the real zero of $x^3 + x + 1 = 0$ correct to hundredths.
Press [**SYMB**] In FUNCTION SYMBOLIC VIEW, in an empty space,
Enter **[x,T,θ][x^y] 3** [+]**[x,T,θ]**[+] **1**
■**[VIEWS]** Highlight **Decimal {{OK}}** The function is graphed.
The function crosses the X axis
between -1 and 0.
Press **[2nd][CALC][2:zero]**
{{MENU}}{{FCN}} Highlight **Root {{OK}}**
The approximation is $x \approx -.682$, correct to 3 places.

Using the Numeric Solver:

ex. Find the real zero of $x^3 + x + 1 = 0$ correct to hundredths.
Press [**LIB**] Highlight **[SOLVE][ENTER]**
In FUNCTION SYMBOLIC VIEW,
Enter **[x,T,θ][x^y] 3** [+]**[x,T,θ]**[+] **1**
■**[VIEWS]** Highlight **Decimal {{OK}}**
The function crosses the X axis
between -1 and 0.
Press [**LIB**] Highlight **Solve**
In SOLVE SYMBOLIC VIEW,
Enter **[x,T,θ][x^y] 3** [+]**[x,T,θ]**[+] **1 {{=}} 0**
[**NUM**] Enter a guess **0 {{OK}}{{SOLVE}}**
The x value is now $x \approx -.682...$ correct to hundredths.
A guess of -1 or -.5 would have worked as well. Try it.

Using the Polynomial Solver:

To find the zeros of a polynomial use POLYROOT.
Find the zeros of $x^3 + x + 1 = 0$
Press [**MATH**] **P** [▷][▽][▽][▽]**[ENTER]**
The entry line reads POLYROOT(

Press ■ [[ ] **1** , **0** , **1** , **1** [ ]][ )][**ENTER**]
The roots $x \approx (.6823,0)$, $(.3411,\pm1.16615)$
are displayed in complex form.

Finding the Intersection of 2 graphs:

ex. Find the intersections of $y = x^2 + 2x - 3$ and $y = \frac{x}{2}$.
Press [**LIB**] highlight **Function**
$Y1$ = [**x**,**T**,θ][x^2][+] **2** [**x**,**T**,θ,**n**][–] **3**
$Y2$ = [**x**,**T**,θ][÷] **2** ■[**VIEWS**] Highlight **Decimal**
You can see the intersections are in the first and third quadrants. TRACE to the intersection in Quad. 1. Using [△] and[▽] arrows you can see the coordinates are not "nice" numbers, so use an INTERSECTION SOLVER.
Press {{**MENU**}}TRACE to the intersection in Quadrant 1.
Press {{**FNC**}} Highlight **Intersection** {{**OK**}}{{**OK**}}
parabola is marked. [**ENTER**]
An approximation to the coordinates is
$x \approx 1.137...$ $y \approx .569...$
Using the same proceedure in Quad 3,
$x \approx -2.367...$ $y \approx -1.319$

13. Finding Maximum and Minimum values of a Function:

The X value is the location of the extrema.
The Y value is the value of the extrema.
Find the minimum value of $f(x) = x^2 + 3x - 1$

Using Graph Solve:

In [**LIB**] highlight **Function** [**SYMB**].
Enter $x^2 + 3x - 1$ in F1(X).
[**x**,**T**,θ][x^2][+] **3** [**x**,**T**,θ][–] **1**
■[**VIEWS**] Highlight **Decimal**
The minimum is off the screen.
Press {{**MENU**}}{{**FCN**}}
Highlight **Extremum** {{**OK**}}
The minimum is $x = -1.5$, $y = -3.25$

14. Statistic Plots:

Press [**LIB**] Highlight **Statistics** [**ENTER**]

Entering Data:

The list editor is displayed.
If necessary, use ■[CLEAR] to clear the lists.
Use the following data:

x	1	2	3	4	5	6	7	8
y	1.1	2.6	3.8	5.1	5.9	7.2	8.2	9

Enter the independent variables, x, in list $C1$.
Highlight the first position in $C1$.
Press **1 [ENTER] 2 [ENTER]** ...
Enter the dependent variables, y, in list $L2$.
Highlight the first position in $L2$.
Press **1.1 [ENTER] 2.6 [ENTER]** ...
Be sure that the type is **{{2VAR□}}**

Constructing a Graph:

Press **■[VIEWS]** Highlight **Auto Scale {{OK}}**
If the regression line is drawn **{{MENU}}{{FIT□}}**
will regraph just the Scatter Plot
Press **[SYMB]**
If necessary clear previous entries, **■[CLEAR]**
Check **S1 {{√CHK}}**

Regression Analysis:

Press **[SYMB]** Highlight **Fit1**:**{{SHOW}}**
The regression equation is displayed in the equation editor.

Regression Plot:

Press **[PLOT]** to display the graph.
The scatter plot.is displayed The regression line will
be shown if {{Fit}} is turned on.

15. Complex Numbers:

Rectangular Complex

To type the number $i = \sqrt{-1}$ **■[A...Z] I** or highlight
and paste from **■[CHARS]**

A complex number can also be entered as
an ordered pair (x, y)

To enter 2 + 3i, press **2** [+] **3 [A...Z] I** or $(2,3)$

To multiply or divide complex numbers, each number
must be entered in a parenthesis.

ex. **(2 + 3i)** + **(3 − 5i)** = $\underline{5-2i}$

Polar Complex

Is not available directly on the HP 38.
You can convert the number to Polar, then enter as $re^{i\theta}$

Complex Functions:

The menu for Complex number calculations is
{MATH] **C** Highlight **Complex** [▸]

16. **Matrices**:

Matrices are *stored* by name, edited and used in matrix arithmetic in the
■**[MATRIX]** menu on the keyboard.
The menu holds 10 matrices named **M1** - **M0**
To enter a matrix, press ■**[MATRIX][EDIT]**
ex. ■**[MATRX][EDIT]**

A 2 × 2 grid is displayed with element a_{11} highlighted.
1 [ENTER] 2 [ENTER][▾] 3 [ENTER] 4 [ENTER]

the matrix M1 = $\begin{bmatrix} 1 & 2 \\ 3 & 4 \end{bmatrix}$ is saved.Press **[HOME]**

Matrix *operations* are displayed with **[MATH][A...Z] M {{OK}}**

Press **R** Highlight **RREF {{OK}}**
The home screen displayes **RREF(**
Press **[A...Z] M 1[ )][ENTER]**
Matrix **M1** is row reduced to [[1,0],[0,1]]
To save it press ■**[ANSWER]{{STO▸}}]**
[A...Z] M 2 [ENTER]

Matrix *arithmetic* +, –, ×,^ is done with the operation keys
on the keyboard.

Matrices may be *entered from the keyboard*:

ex. Find the solution to the system of equations:

$$\begin{aligned} -2x_1 + 3x_2 - 2x_3 &= 16 \\ -5x_1 + 3x_2 - 5x_3 &= 22 \\ x_1 \qquad\quad + x_3 &= -2 \end{aligned}$$

Enter the matrix $\begin{bmatrix} -2 & 3 & -2 & 16 \\ -5 & 3 & -5 & 22 \\ 1 & 0 & 1 & -2 \end{bmatrix}$

Use [(-*x*)] for the negative numbers
[[–2, 3, –2, 16] , [–5, 3, –5, 22] , [1, 0, 1, –2]]
[STO▸][A...Z] M 2 [ENTER]

Find the reduced row echelon form:

[MATH] M Highlight **Matrix** [▷] **R**
Highlight **RREF {{OK}}**
On the home screen **RREF(** is displayed
Press **[A...Z] M 2** [)]**[ENTER]**

Find the inverse of a matrix:

Enter and store: $\begin{bmatrix} 1 & 1 & 2 & 1 \\ -1 & 1 & 0 & 1 \\ 2 & 1 & 1 & 0 \\ 1 & 3 & 1 & 0 \end{bmatrix}$

Use [(–*x*)] for a negative number.
[[**1**, **1**, **2**, **1**], [**–1**, **1**, **0**, **1**], [**2**, **1**, **1**, **0**], [**1**, **3**, **1**, **0**]]
{{STO▸}}[A...Z] M 3 [ENTER]
{MATH] M {{OK}} I Highlight **Inverse {{OK}}**
On home screen enter **[A...Z] M 3**) **[ENTER]**
{{STO▸}} [A...Z] M 4
Use **[▲] {{SHOW}}** to view the rest of the matrix.

To check the inverse:

[A...Z] M 3 [×]**[A...Z] M 4 [ENTER]**
The answer is the Identity of order 4.

To find the determinent of matrix A:

[MATH] M Highlight **MATRIX]**[▷] **D**
Highlight **DET {{OK}}**
On the home screen enter **[A...Z] M 3 [ENTER]**
The answer is 6.

Chapter 9 – Using the SHARP EL9600c Graphing Calculator.

1. Notation:

Keystrokes, except for numbers and commas are in bold brackets **[MATH]**
For a second function $\sqrt{\ }$, press **[2nd]** [$\sqrt{\ }$] the second functions are printed in yellow on the left over the keys.
For the letter A, press **[ALPHA][A]** the ALPHA symbols are printed in blue on the right over the keys.
For functions in the screen menus 2^x, press **[MATH][A: CAL][02: 2^x][X/θ/T/*n*]**
or **[MATH]** highlight **[CALC]** [▷][▷]**[ENTER]** or use the touch pen press **[MATH]** touch **A** then touch **02** then **2^x** then [±×÷]
2^ is displayed on the home screen.
In this text 2^x is keystroked **[MATH][A][02: 2^x]**
Calculator results are underlined.

2. Basics:

To adjust the screen contrast: Press **[2ndF][OPTION]**
then [△] to increase the contrast or [▽] to decrease the contrast.
Press **[2ndF][QUIT]** to leave the menu.
Mode settings: Press **[2ndF][SET UP]** the current settings are in the box on the right side of the screen. They should read

Rad
FloatPt
9
Rect
Decimal(Real)
One line

To change a MODE setting, press the letter for the setting, the choices appear in the box on the right. Press the number for the choice
The updated choices are in the right box. You may use the touch pen or you may enter from the keyboard.
Use [±×÷] or **[2ndF][QUIT]** to go to the home screen.

3. Arithmetic Computations:

Order of Operations: The order is Algebraic, operations are performed from left to right. First exponentiation, then multiplication and division, then addition and subtraction. Parenthesis must be used to change the algebraic order. Parenthesis must be used around the numerator and denominator in Algebraic Fractions.

Minus sign: Use [(–)] for negative, [–] for subtraction.

Fractions: to enter fractions use [a/b]

To change a decimal to a fraction use **[2ndF][SET UP]**

Set the mode to **[F][3:Improp(Real)]**[±]

To change a fraction to a decimal use **[2ndF][SET UP]**

Set the mode to **[F][1:Decimal(Real)]**[±]

To calculate: press **[ENTER]**

(Note) $\times$ is displayed as $\star$ and $\div$ is displayed as /

ex. $2 \times 3 + 4 \times 5$ **[ENTER]** 26

$2 \times (3 + 4) \times 5$ **[ENTER]** 70

$3 + 4 \div 2$ **[ENTER]** 5

$(3 + 4) \div 2$ **[ENTER]** 3.5

1 [a/b] 3 **[ENTER]** .333333333

[2ndF][SET UP][F][3:Improp][±] 1⌐3

4. **Scientific Notation**:

From the keyboard use number between 1 and 10 **[2nd][10^x]** exponent.
The exponent must be a number between -99 and 99.
To Display answers in Scientific Notation change the MODE menu entry.

[2ndF][SET UP][C][3:Sci]

You may want to change the number of decimal places displayed. To display 3 places, change to FIX 3 press

[2ndF][SET UP][D][3]

To return to Normal:

[2ndF][SET UP][C][1] and **[D][9]**

5. **Editing**:

The Sharp has 2 Editors. One is the Normal Graphing Calculator Entry, the other is an equation editor that writes functions as how you see them, "pretty print". This manual will use the normal One Line Editor.

Before typing ***[ENTER]***

A blinking cursor determines the current position on the screen.
The arrow keys move the cursor around the screen.

To change an entry move the blinking cursor to the entry, then type the new entry. It will replace the old entry.

To delete a symbol, move to the symbol, press **[DEL]**

To insert a symbol, move to the symbol after the insertion point, press **[2ndF][INS]** then type the new text.

To erase the previous symbol, press the backspace key **[BS]**

After typing **[ENTER]**

Press **[2ndF][ENTRY]** the last command is recalled, edit it as explained above in *Before typing ENTER*

You may continue to press **[2ndF][ENTRY]** to go back to other commands.

To clear the home screen press **[CL]**

To exit a MENU, press **[2ndF][QUIT]** or enter another MENU.

To erase a function from the Y = screen, highlight any symbol in the function, press **[CL]**

6. Entering Algebraic Functions:

From the Y = screen, use the arrow keys to move to an empty position. Enter the function. It is now available to use as a graph or a table. This menu holds 10 functions.

On the home screen, enter the function. When you press **[ENTER]** it will be evaluated with the values of the variables that are stored in Memory.

When you enter built in functions, parenthesis need to be included Complete the input variable, then type the right parenthesis.

The key **[X/θ/T/*n*]** prints x in Function mode, T in parametric mode, θ in polar mode and n in sequential mode.

Special Functions:

Absolute value of x: press **[MATH][B:NUM][1:abs][X/θ/T/*n*]**

Powers, x^5: press **[X/θ/T/*n*][^] 5**

3^x: press **3 [^][X/θ/T/*n*]**

e^x: press **[2ndF]**[e^x]**[X/θ/T/*n*]** the ^ is forced.

Roots, $\sqrt[5]{x}$: press **5 [2ndF][a√‾][X/θ/T/*n*]**or **[X/θ/T/*n*][^][(]1 [÷] 5 [)]**

Natural log, $ln(x+5)$: press [ln][(]**[X/θ/T/*n*]**[+] **5 [)]**

Conjugate Pairs, $\pm\sqrt{4-x^2}$: press [{] **1,-1** [}]**[2nd]**[√‾][(] **4** [–] **[X/θ/T/*n*][x²][)]**

Factorial, 5!: press **5 [MATH][C][4:!]**

Trig function, $tan(3x^2)$: press **[tan]**[(] **3 [X/θ/T/*n*][x²][)]**

Inverse trig function, $arcsine(x)$: press **[2ndF][sin⁻¹][X/θ/T/*n*][)]**

Combinations, ${}_5C_2$: press **5 [MATH][C][3:**${}_nC_2$**] 2**

7. Evaluating a function: (checking your answer)

Variables are represented as single alphabetic characters, A through Z, and θ. Numbers are stored in these positions using the command **[STO]** an arrow ⇒ is shown on the home screen.

When an algebraic expression is entered, the variables are replaced with the stored constants and a number is displayed.

To evaluate a function on the home screen use the key **[ALPHA]** [:] to connect commands.

ex. To evaluate $x^2 + 3x - 1$ for $x = 2$,
Press **2 [STO][X/θ/T/*n*][ALPHA]**[:]
[X/θ/T/*n*][x²][+] **3 [X/θ/T/*n*]**[–] **1**
[ENTER]
the result is 9.

If the function is stored in the Y = menu it may be recalled from the VARS menu.

ex. Enter the function $x^2 + 3x - 1$ in the Y = menu as Y1.
Press the blue key **[Y=]**, press **[CL]**, then
[X/θ/T/*n*][x²][+] **3 [X/θ/T/*n*]**[–] **1 [ENTER][2nd][QUIT]**
Store the number 2 in memory x , **2 [STO▸][X/θ/T/*n*][ENTER]**
Press **[VARS][ENTER][1:Y1][ENTER]**
the result is 9.

All expressions containing an x will be evaluated with $x = 2$ until this number stored in the x box is replaced with another number.

8. Angles:

The type of measure for angles is set in the MODE menu.
Set **[2ndF][SET UP]B [2:Rad]** for **Radians** or **[2ndF]**
[SET UP]B for **Degrees [ENTER]**
After the MODE is set, use the operations in **[2nd][ANGLE]**
to change the units.

To enter an angle in Degrees while in **Radian** MODE:
ex. Press **[SIN] 30 [MATH] E [1:°]**[)]**[ENTER]** .5

To enter an angle in Radians while in **Degree** MODE:
ex. Press **[SIN]** π ÷ **6 [MATH][4:ʳ]**[)]**[ENTER]** .5

To change Rectangular Coordinates to Polar (Degree Mode):
ex. Press **[MATH] D** [**3:xy→r(**] **1,1** [)]**[ENTER]**
1.4142... is the radius in polar form.
Press **[MATH] D** [**4:xy→** θ(] **1,1** [)]**[ENTER]**
45 is the angle in **Degrees.**

To change Polar Coordinates (Degree Mode) to Rectangular:
ex. Press **[MATH] D [5:rθ →x(]** $\sqrt{2}$**,45** [)]**[ENTER]** 1
is the x coordinate in rectangular coordinates.
Press **[MATH] D** [**6:rθ →y(**] $\sqrt{2}$**,45** [)]**[ENTER]** 1
If the MODE is Radians, enter the angle in Radians.

9. **Building Tables**:

The Sharp has a built in TABLE function used to evaluate functions for different values of x. The algebraic expression to be evaluated is stored in the Y = menu. In the TABLE SET menu, set the initial value for x, and the step size between entries. Use AUTO for input or set USER to enter a random set of numbers for x.

ex. Enter $x^2 + 3x - 1$ in Y1 in the Y = Menu.
Press **[2ndF][TBLSET]** set the following values:
[(–)] **4 [ENTER]** for TblStart
1 [ENTER] forTblStep
Press **[2ndF][QUIT]** when finished.
Press **[TABLE]** to view the table.

Use up and down arrows to see more values in the table.

10. **Building Lists**:

Numbers can be stored in Lists either on the home screen or in the LIST TABLE

Lists may be named (stored) in the data set
[L1],**[L2]**,**[L3]**,**[L4]**,**[L5]**,**[L6]**using the **[2ndF]** keys.

You may store 10 such data sets.

ex. Enter the numbers 1, 3, 5, 7 in list **L1**
[2ndF][{] 1 , 3 , 5 , 7 [2ndF][}][STO][2ndF][L1][ENTER]

To display L1 on the home screen, press **[2ndF][L1][ENTER]**

Arithmetic operations are done using the +,–,×,÷,^ keys.

Other operations are done from the **[2nd][LIST]** **OPE** and **MATH** menus.

Special Lists:

OPE [**5:seq(**] expression, begin, end increment)
Generate a sequence of numbers using a rule for expression.

ex. **[2nd][LIST] A [5:seq]**
[5:seq(][x,T,θ,n][x²] , 1 , 5 , 2][)][ENTER]
square the numbers starting with 1, adding 2, ending at 5.
The result is { 1, 9, 25}

OPS [**6:cumul(**] (list) returns a list of cumulative sums starting with the first entry. For L1, { 1 , 4 , 9 , 16 }

MATH [**5:sum(**] (list) returns the sum of all the numbers in the list. For L1, the sum is 16.

OPS [**7:df list(**] (list) returns a list of the differences of consecutive terms. For L1, the list is { 2 , 2 , 2 }

This list is 1 shorter than the original list.

11. Graphing Functions:

The function to be graphed must be entered as one of the 10 functions
in the Y = menu. Enter $x^2 + 3x - 1$ in Y1.

The function must be turned on to graph, so,
the = must be highlighted.
To turn the function on or off,
highlight the = and press [**ENTER**]
Each ENTER reverses the current state.

To graph the function press the blue key [**GRAPH**]
The function is graphed on the current window settings.

The ZOOM menu has choices to zoom in or out
and for built in window sizes.

Press the blue key [**ZOOM**] press [**7:Dec**]
The window is set to a Friendly window
Each pixel is 0.1 units, and the graph is its
true shape. Unit size on X axis = unit size on Y axis.
Any multiple or translation of these settings
also makes a Friendly window.

Press [**ZOOM**][**5:Default**] This is the Standard window, it sets
X to [-10,10] and Y to [-10,10]. The units on the X and Y axis are not the
same size, so graphs are not their true shape.

Press [**ZOOM**][**6:Square**] Now the graph is its true shape again.
The window settings for the Y axis are left the same, and the X settings
have been recalculated so that the units on the X and Y axis are the
same. This is called a Square window. It usually is not Friendly.

Press [**ZOOM**][**1:Auto**] This is the AutoScale function.
After the Xmax, Xmin are set, it finds a Y-range to show a
Complete Graph. The Y settings often need adjustment to
get a clear picture. With ZoomFit a Friendly Window stays Friendly.

Press the blue key [**WINDOW**] It displays the current settings for the X and
Y ranges. These values can be changed by using the arrow keys to
highlight the line, then typing the new value.
Use [**ENTER**] after each change.
(Remember to use (–) for a negative number).
The current values can be multiplied by a constant, or a constant
can be added to an entry and the calculator will do the arithmetic.
Press [**GRAPH**] to return to the graph.

Press the blue key [**TRACE**] A blinking star cursor appears on the Y axis.

The left and right arrow keys move this cursor along the function graph.
The coordinates that are displayed are points that satisfy the function equation. The Y values are computed from the X value of the pixel.
When more than one function is plotted, the up and down arrows move the cursor vertically between the different plots. These coordinates are only approximations unless the graph is on a Friendly window and the coordinate is a rational number.

To graph a split function: $f(x) = \left\{ \begin{array}{ll} x+7, & x \le -5 \\ 4-x, & x > -5 \end{array} \right\}$

Press [Y=] then enter the function in 2 empty spaces.
$Y1 = (x+7) \div (x \le -5)$ $Y2 = (4-x) \div (x > -5)$
$Y1 =$ [(][**x,T,θ,n**][+] **7** [)][÷]
[(][**x,T,θ,n**][**MATH**] **F** [**6**:≤][(–)] **5** [)]
$Y2 =$ [(] **4** [–][**x,T,θ,n**][)][÷][(][**x,T,θ,n**]
[**MATH**]**F** [**3**:>][(–)] **5** [)]
Graph using **[ZOOM][5:Default]**

To graph Parametric Equations: press **[2ndF][SET UP]**
E [2:Param] [ENTER]
Press [Y=] the menu now reads **X1T** and **Y1T**
You may want to change the window settings.
Now X and Y are both functions of the independent variable T.
The settings for Tmin and Tmax determine how much of the graph is plotted.
Tstep determines how many points are plotted.
Enter the functions and press **[GRAPH]**

ex. Write the function $y = x^2 + 3x - 1$ in parametric form.
Let $x = T$, $y = T^2 + 3T - 1$.
Enter **X1T = T**
Y1T = T^2+3T – 1
Press **[WINDOW]**
Set **Tmin** = [(-)] **5**, **Tmax = 5.**
Press **[ZOOM] A [7:Dec]** The graph is displayed.
Use **[TRACE]** This is **not** a Friendly Window

To graph implicit functions: easily,
Write the equation in Parametric form.

ex. Write the equation $x^2 + y^2 = 4$
in parametric form: $x = 2\cos(t)$, $y = 2\sin(t)$.
Enter **X1T= 2**cos**(T), Y1T= 2**sin**(T)**
[ZOOM] A [7:Dec] The graph is a circle.

Remember to change back to Function Mode.

To graph *Polar Equations*: press **[2ndF][SET UP]**

E [3:Polar] [ENTER]

Press **[Y=]**, Enter **r1** $= \theta$, where r = f(θ).

Polar is a special case of Parametric, so

Set $\theta \min = 0$ and $\theta \max = 4\pi$ **[ZOOM]A [5:Default]**

Remember to change back to Function Mode.

12. Solving Equations:

Write the equation in the form $f(x) = 0$

Using Trace and Zoom will give you an approximate answer.

Use either a Zoom box or use Zoom In.

To Set the ZOOM Factors:

Press **[ZOOM] B [ENTER] 4 [ENTER] 4 [ENTER] [2ndF][QUIT]**

ex. Find the real zero of $x^3 + x + 1 = 0$
correct to hundredths. (Use both Zoom factors = 4)

Use Zoom In:

Press **[Y=]** in an empty space

Enter **[x,T,θ,n][a^b] 3** [+]**[x,T,θ,n]**[+] **1**

[ZOOM] A [7:Dec]

The function crosses the X axis
between -1 and 0.

Press **[WINDOW]**

Set the Xscl = 0.01 **[GRAPH]**

Press **[TRACE]**

Move the cursor near the root.

[ZOOM] A [3:In] Repeat

[TRACE][ZOOM] A [3:In] Repeat

[TRACE][ZOOM] A [3:In]

Now the tick marks are visible on the X-axis.

[TRACE]

For $x \approx -.6828125, y \approx -.0011622$

For $x \approx -.68125, y \approx .00258081.$

The value of the zero is $x \approx -.68...$ correct to hundredths.

Note that these 2 values of y lay on opposite sides of the X axis.
(These figures depend on the value of the Zoom Factors, these are set at 4).

Use Zoom Box:

Press **[Y=]**, in an empty space enter **[x,T,θ,n][a^b] 3** [+]
[x,T,θ,n][+] **1 [ZOOM] A [7:Dec]**

Note that these 2 values of y lay on opposite sides of the X axis.
(These figures depend on the value of the Zoom Factors, these are set at 4).

Use Zoom Box:

Press [Y=], in an empty space enter **[x,T,θ,n][a^b] 3** [+] **[x,T,θ,n]**[+] **1 [ZOOM] A** [7: **Dec]**
You can see that the function crosses the X axis between -1 and 0.
Press [**WINDOW**] and set the Xscl = 0.01 [**GRAPH**]
Move the free cursor with the arrow keys until it is above and to the left of the zero. Press **[ZOOM][2: Box][ENTER]**
Note that the cursor is now a blinking box.
Use [▷], then [▽] to draw a box with the zero inside **[ENTER]**
Repeat until the tick marks on the X axis are clearly visible.
Trace to approximate the value $x \approx -.68...$

Using the Graph Solve feature:

ex. Find the real zero of $x^3 + x + 1 = 0$ correct to hundredths.
Press [Y=] in an empty space enter **[x,T,θ,n]**[^] **3** [+] **[x,T,θ,n]**[+] **1 [ZOOM] A** [7: **Dec]**
You can see that the function crosses the X axis between -1 and 0.
Press **[2nd][CALC][5: X-Incpt]**
The approximation is $x \approx -.682...$ correct to 3 places.

Using the Polynomial Solver.

The POLYNOMIAL SOLVER finds all the zeros of a 2nd or 3rd degree polynomial.

ex. Find all the zeros of $x^3 + x + 1$
Press **[2nd][TOOL][C] 3**
Enter the coefficients.
1 [ENTER][ENTER] 1 [ENTER] 1 [ENTER]
[2ndF][EXE] All of the zeros are displayed.
$x_1 \approx -.68233$, $x_2 \approx .3412 + 1,1615i$, $x_3 \approx .3412 - 1.1615i$

Finding the Intersection of 2 graphs:

ex. Find the intersections of $y = x^2 + 2x - 3$ and $y = \frac{x}{2}$.
Press [Y=] in empty spaces enter
$Y1$ = **[x,T,θ,n]**[x^2][+] **2 [x,T,θ,n]**[–] **3**
$Y2$ = **[x,T,θ,n]**[÷] **2 [ZOOM] A** [7: **Dec]**
You can see the intersections are in the first and third quadrants. [**TRACE**] to the intersection in Quad. 1. Using [△] and[▽] arrows you can see the coordinates are not "nice" numbers, so use an

INTERSECTION SOLVER.
Press **[2nd][CALC]**
Move the cursor
near the intersection in Quad 1, [[**2**: **Intsct**]
An approximation to the coordinates is
$x \approx 1.137...\quad y \approx .569...$
Move the cursor
near the intersection in Quad 3, [[**2**: **Intsct**]
$x \approx -2.367...\quad y \approx -1.319...$

13. **Finding Maximum and Minimum values of a Function**:

The X value is the location of the extrema.

The Y value is the value of the extrema.

Find the minimum value of $f(x) = x^2 + 3x - 1$

Using Graph Solve:

Press **[Y=]**, in an empty space enter
[x,T,θ,n][x^2][+] 3 [x,T,θ,n][–] 1
[ZOOM] A [7: Dec]
The minimum is off the screen.
Press **[WINDOW]**
Change Ymin to -5.1 and Ymax to 1.1
[GRAPH]
[2nd][CALC][3: minimum]
The minimum is $y = -3.25$

14. **Statistic Plots**:

Entering Data:

Press **[STAT] 1** to display the list editor.
Clear all data from each list by pressing **[▵][CLEAR][ENTER]**
Use the following data:

x	1	2	3	4	5	6	7	8
y	1.1	2.6	3.8	5.1	5.9	7.2	8.2	9

Enter the independent variables, x, in list $L1$.
Highlight the first position in $L1$.
Press **1 [ENTER] 2 [ENTER]** ...
Enter the dependent variables, y, in list $L2$.
Highlight the first position in $L2$.
Press **1.1 [ENTER] 2.6 [ENTER]** ...

Constructing a Graph:

Press [Y =] and turn off or clear any functions.
Press **[2nd][STAT PLOT] 1** to select Plot 1
Press **[ENTER]** to turn ON Plot 1.
Make the following selections, highlight, press **[ENTER]**

PLOT1 on
DATA XY
ListX:L1
ListY:L2
Highlight GRAPH, Press **[2ndF][STAT PLOT]**
[▷] pick the mark for the plot
[3: Scatter∘]

Press **[ZOOM] A [9: Stat]** to see the Scatter Plot

Regression Analysis:
and
Regression Plot:

A regression analysis is needed before a Plot can be drawn.
On the home screen, press **[CL]**
[STAT] D [02: Rg_ax+b][ENTER]
[(][2ndF][L1],[2ndF][L2],[VARS] A [ENTER] A 1 [)]{ENTER]
The current regression equation is displayed on the screen and copied to Y1..
Press **[GRAPH]**
Watch the regression line plot through the scatter plot.

15. Complex Numbers:

Rectangular Complex

The number $i = \sqrt{-1}$ is **[2ndF] i** (over the period).
To enter 2 + 3i, press **2 + 3 [2nd][i][ENTER]**
To multiply or divide complex numbers, each number must be entered in a parenthesis.

ex. **(2 + 3i)** + **(3 – 5i)** = $\underline{5 - 2i}$

Polar Complex

The number a + bi is written $re^{i\theta}$

$r = \sqrt{a^2 + b^2} \quad \theta = \arctan(\frac{y}{x})$

Coversions can be made between polar and rectangular, then the number can be changed into Polar complex or Rectangular complex form.

Complex Functions:

Press **[MATH] H** to display the list of functions and operations for complex numbers.

16. Matrices:

Matrices are *stored* by name, edited and used in matrix arithmetic in the MATRX menu on the keyboard.

The menu holds 10 matrices named A - J.

To enter a matrix, press **[MATRX][▹][▹][ENTER]** (EDIT)

Enter the dimensions, rows × columns then fill in the matrix using **[ENTER]** after each entry.

ex. **[MATRX][▹][▹][ENTER] 2 [ENTER] 2 [ENTER]**

A 2 × 2 grid is displayed with element a_{11} highlighted.

1 [ENTER] 2 [ENTER] 3 [ENTER] 4 [ENTER]

[2nd][QUIT] the matrix [A] = $\begin{bmatrix} 1 & 2 \\ 3 & 4 \end{bmatrix}$ is saved.

Matrix *operations* are displayed with **[MATRX][▹]** (MATH)

From this screen press **[ALPHA] B**

The home screen displayes **rref(**

Press **[MATRX][1**:[A]]**[)][ENTER]**

Matrix [A] is row reduced to $\begin{bmatrix} 1 & 0 \\ 0 & 1 \end{bmatrix}$

To save it press **[2nd][ANS][STO▸][MATRX][2**:B]**[ENTER]**

The matrix name [A] or [B] must be pasted onto the home screen from the MATRIX NAMES Menu.

Matrix *arithmetic* +, −, ×,^ is done with the blue operation keys on the keyboard.

Matrix *Row Opreations* are done from the menu **[MATRX]** [▹] MATH submenu,

Press [△]. The row operations are **C**:,**D**:, **E**:, **F**:.

See the Sharp 9600 manual for details.

Matrices may be *entered from the keyboard*:

ex. Find the solution to the system of equations:

$$\begin{aligned} -2x_1 + 3x_2 - 2x_3 &= 16 \\ -5x_1 + 3x_2 - 5x_3 &= 22 \\ x_1 \qquad\quad + x_3 &= -2 \end{aligned}$$

Enter the matrix $\begin{bmatrix} -2 & 3 & -2 & 16 \\ -5 & 3 & -5 & 22 \\ 1 & 0 & 1 & -2 \end{bmatrix}$

Use [(–)] for the negative numbers
[2nd][[ ][2nd][[ ] -2 , 3 , -2 , 16 [2nd][ ]]
[2nd][[ ] -5 , 3 , -5 , 22 [2nd][ ]]
[2nd][[ ] 1 , 0 , 1 , -2 [2nd][ ]]
[2nd][ ]][ENTER]
[STO▸][MATRX][ENTER][ENTER]

Find the reduced row echelon form:

[MATRX][▹][ALPHA][B][MATRX][1:[A]][ENTER]

Find the inverse of a matrix:

Enter and store: $\begin{bmatrix} 1 & 1 & 2 & 1 \\ -1 & 1 & 0 & 1 \\ 2 & 1 & 1 & 0 \\ 1 & 3 & 1 & 0 \end{bmatrix}$

Use [(–)] for a negative number.
[2nd][[ ][2nd][[ ]1 , 1 , 2 , 1 [2nd][ ]]
[2nd][[ ] -1 , 1 , 0 , 1[2nd][ ]]
[2nd][[ ] 2 , 1 , 1 , 0 [2nd][ ]]
[2nd][[ ] 1 , 3 , 1 , 0 [2nd][ ]]
[2nd][ ]][ENTER]
[STO▸][MATRX][1:[A]][ENTER]
[MATRX][1:[A]][x^{-1}][ENTER]
[STO▸][MATRX][2:[B]]
[ENTER][MATH][1:▸Frac][ENTER]
Use [▹] to view the rest of the matrix.

To check the inverse:

[MATRX][1:[A]][×][MATRX][2:[B]][ENTER]
The answer is the Identity of order 4.

To find the determinent of matrix A:

[MATRX][▹][1:det(][MATRX][1:[A]][)][ENTER]

The answer is 6.